IMPROVING ENVIRONMENTAL IMPACT ASSESSMENT FOR BETTER INTEGRATED COASTAL ZONE MANAGEMENT

Promoter:	Prof. dr. W. van Vierssen Professor of Aquatic Ecology Wageningen University, The Netherlands
Promoter:	Prof. dr. ir. F.R. Rijsberman Director General International Water Management Institute, Sri Lanka
Co-promoter:	Dr. W.J.A.M. Douven Doctor in Environmental Systems Analysis UNESCO-IHE Institute for Water Education, The Netherlands

Awarding Committee:	Prof. dr. ir. K.J. Beek International Institute for Geo-information Science and Earth Observation, The Netherlands
	Dr. ir. I. Soesilo University of Indonesia, Indonesia
	Prof. dr. ir. A.P.J. Mol Wageningen University, The Netherlands
	Prof. P. Burbridge University of Newcastle, United Kingdom

IMPROVING ENVIRONMENTAL IMPACT ASSESSMENT FOR BETTER INTEGRATED COASTAL ZONE MANAGEMENT

Dwi Abad Tiwi

CRC Press
Taylor & Francis Group
Boca Raton London New York

CRC Press is an imprint of the
Taylor & Francis Group, an **informa** business
A BALKEMA BOOK

DISSERTATION
Submitted in fulfilment of the requirements of
the Academic Board of Wageningen University and
the Academic Board of the UNESCO-IHE Institute for Water Education
for the Degree of DOCTOR
to be defended in public
on Monday, 19 January 2004 at 15:00 h in Delft, The Netherlands

Published by:
CRCPress/Balkema
P.O. Box 447, 2300 AK Leiden, The Netherlands
e-mail: Pub.NL@taylorandfrancis.com www.crcpress.com
www.taylorandfrancis.com

First issued in hardback 2018

ISBN 13: 978-1-138-44078-4 (hbk)
ISBN 13: 978-90-5809-654-8 (pbk)

Visit the Taylor & Francis Web site at
http://www.taylorandfrancis.com

and the CRC Press Web site at
http://www.crcpress.com

TABLE OF CONTENTS

	Page
List of Tables	i
List of Figures	iii
Frequently Used Glossary	v
SUMMARY	1
PART I. INTRODUCTION	
1. INTRODUCTION	9
1.1. Background	9
1.2. Structure of the EIA research to support ICZM in the Banten Bay area	15
1.3. Structure and Contents of the Ph.D Thesis	16
PART II. CONCEPT AND PRE-ANALYSIS	
2. INTEGRATED COASTAL ZONE MANAGEMENT AND ENVIRONMENTAL IMPACT ASSESSMENT	21
2.1. Introduction	21
2.2. Integrated Coastal Zone Management (ICZM)	21
2.2.1. Decision-making in ICZM	21
2.2.2. Environmental information and stakeholder participation in ICZM	23
2.3. ICZM and Environmental Impact Assessment (EIA)	27
2.3.1. ICZM and EIA	27
2.3.2. Environmental information in EIA	28
2.3.3. Stakeholder participation in EIA	29
2.4. Effectiveness of EIA to Improve ICZM	32
2.4.1. Basic principles for effective EIA practice	32
2.4.2. Problems hindering effective use of EIA	33
2.4.3. Application of effective EIA	33
2.4.4. Effective use of coastal EIA to improve ICZM	34
2.5. Conclusions	35
3. ICZM AND EIA IN INDONESIA	37
3.1. Introduction	37
3.2. Indonesian Coastal Planning and Management	37
3.2.1. Planning system, laws and regulations	37
3.2.2. Position and role of government	38
3.3. Recent Changes in Indonesian Policies and in Roles of District Government	40
3.3.1. Changes in Indonesian policies	40

3.3.2. Influence of the institutional changes on the governance system and the EIA process 40
3.4. Management Problems in Indonesian ICZM and EIA 44
3.4.1. Management problems in the Indonesian ICZM 44
3.4.2. Management problems in the Indonesian EIA system 45
3.5. Preliminary Research in the Banten Bay Area 49
3.5.1. Study area 49
3.5.2. Government strategies in coastal management 50
3.5.3. Development procedures in the Banten Bay area: two cases 50
3.6. Conclusions 57

4. METHODOLOGIES OF THE RESEARCH
4.1. Introduction 59
4.2. Problem Analysis 59
4.3. Central Problem Statements 62
4.4. Research Design: Use of Environmental Information in EIA Process 63
4.4.1. Objectives and research questions 63
4.4.2. Assessment of EIA guidelines 65
4.4.3. Assessment of coastal EISs 68
4.4.4. Assessment of environmental information sources to draft and review EIS 72
4.4.5. Observation of the coastal EIS review meetings 74
4.4.6. Assessment of environmental monitoring 76
4.4.7. Assessment of perceptions of technical team on the EIA process 77
4.5. Research Design: Stakeholder Participation in the coastal EIA Process 78
4.5.1. Objectives and research questions 78
4.5.2. Assessment of stakeholder participation in the EIA process based on perceptions of technical teams 80
4.5.3. Observation of stakeholder participation in the coastal EIS review meetings 81
4.5.4. Assessment of stakeholder participation in the pre-coastal EIS compilation 82

PART III. ANALYSIS OF THE CASE STUDIES

5. USE OF ENVIRONMENTAL INFORMATION IN THE COASTAL EIA PROCESS 87
5.1. Introduction 87
5.2. Environmental Information in the EIA Guidelines 87
5.2.1. EIA technical guideline of the Industry sector 87
5.2.2. EIA technical guideline of the Sea Communications sector 88
5.2.3. Indonesian EIA general guideline 88
5.3. Environmental Information Described in the Coastal EISs 90
5.3.1. Quality of the coastal EISs reviewed based on all reports and sectoral reports 90
5.3.2. Quality of the coastal EISs reviewed based on sectors and two dominant cases per sector 92
5.4. Information Sources Used by Consultants to Draft EIS 93
5.5. Information Sources Used by the Technical Teams and the District Governments to Review EIS 96

5.5.1. Perceptions of the technical team respondents 97
5.5.2. Perceptions of the district government respondents 99
5.6. Environmental Information Discussed in the Coastal EIS Review Meetings 102
5.6.1. Review of the coastal EIS contents in the meetings 102
5.6.2. Discussions by participants in the meetings 104
5.7. Monitoring of Environmental Parameters 106
5.7.1. Parameters for environmental monitoring 106
5.7.2. Parameters for environmental impact monitoring 108
5.8. Environmental Information Based on Perceptions of Technical Team 111
5.8.1. Use of the EIA technical guidelines 111
5.8.2. Environmental information in the EIA technical guidelines 112
5.8.3. Implementation law on regional autonomy and its effect on the EIA Process in 1999 113
5.9. Conclusions 114

6. STAKEHOLDER PARTICIPATION IN THE COASTAL EIA PROCESS 121
6.1. Introduction 121
6.2. Perceptions of Technical Teams on Stakeholder Participation 121
6.3. Stakeholder Participation in the Coastal EIS Review Meetings 124
6.3.1. Participation of stakeholders during the meetings 125
6.3.2. Inputs from stakeholders to the meetings 126
6.3.3. Objectivity of the coastal EIA review meetings 128
6.4. Stakeholder Participation in the Pre-Coastal EIS Compilation 129
6.4.1. Participation of stakeholders in project approval 130
6.4.2. Perception of stakeholders on communication and information input to the project 132
6.4.3. Perception of stakeholders on access to information 134
6.5. Conclusions 136

7. DISCUSSION 141
7.1. Introduction 141
7.2. Limited Availability of Priority Information 141
7.3. Limited Information Sharing 143
7.4. Poor Information Management 143
7.5. Limited Stakeholder Participation in the Coastal EIA Process 145
7.6. Lack of Empowerment of Stakeholders 146
7.7. Future Direction of Stakeholder Participation 147
7.8. Improving Use of Environmental Information and Stakeholder Participation 149
7.8.1. Promoting better use of environmental information 149
7.8.2. Promoting stakeholder participation 150
7.8.3. Promoting effective EIA, supporting decentralization process, and building local capacity 151

8. RECOMMENDATIONS TO IMPROVE THE COASTAL EIA PROCESS 153
8.1. Introduction 153
8.2. Improved Use of Environmental Information and Improvement of Stakeholder Participation 153
8.2.1. Improved use of environmental information 153

8.2.2. Improvement of stakeholder participation 159
8.2.3. Improvement of external factors 162
8.3. Instruments to Implement Some of the Recommendations 164
8.3.1. Banten Bay Information System (BBIS) 164
8.3.2. EIS review criteria 166
8.3.3. Measures for empowering stakeholders 167
8.4. Conclusions 169

PART IV. CONCLUSIONS

9. CONCLUSIONS 173
9.1. Introduction 173
9.2. Reflection on Central Problem Statement 173
9.3. Reflection on Conceptualisation and Application 174
9.4. Discussion of the Research 175
9.5. Suggestions for Further Research 177

References 179
Samenvatting 191
Acknowledgements 197
Curriculum Vitae 199
Appendix I : Proposed Addition of Indicators for EIA Technical Guideline of the Sea Communications Sector 201
Appendix II : Proposed Addition of Indicators for EIA Technical Guideline of the Industry Sector 205
Appendix III : Quality of 10 coastal EISs reviewed: Summary 209
Appendix IV : Review Criteria for ToR for production of coastal EIS 211
Appendix V : EIS Review Criteria 215
Appendix VI : List of Respondents and Experts 221
Appendix VII: A Quick Tour Around Banten Bay Information System 223

LIST OF TABLES

		Page
3.1.	Changes in Indonesian policies since 1982 and their influence on the governance systems and the EIA process	42
3.2	Summary of management problems related to coastal zones	45
3.3.	Stages of EIA process and need for environmental information (based on government regulation no. PP 51/1993)	47
3.4.	Stages of the EIA process (based on decree no. PP 51/1993) and stakeholder participation (based on decree no 08/2000)	49
4.1	Summary of the methodologies used to assess the use of environmental information in the coastal EIA process	66
4.2	The coastal EISs reviewed in the research	69
4.3	Ranks of scale of importance for assessment areas of the coastal EISs reviewed	71
4.4	Scale of ranks for assessment areas of the coastal EIS reviewed	71
4.5	Total weighted ranks for assessment areas of the coastal EISs reviewed	72
4.6	Distribution of questionnaire and returns in the research	74
4.7	Coastal EIS review meetings observed in the research	75
4.8	Summary of the methodologies used in the assessment of stakeholder participation in the coastal EIA process	79
4.9	Distribution of focus groups of respondents in the research	84
5.1	Summary of the results of the EIA guidelines reviewed by experts	89
5.2	Information sources mostly used by the consultant respondents to draft EIS	95
5.3	Information sources mostly used by the technical team respondents to review EIS	98
5.4	Information sources mostly used by the district government respondents to review EIS	101
5.5	Summary of observations of the 4 coastal EIS review meetings	105
5.6	Environmental monitoring indicated by the district government of Serang	107
5.7	Constraints in environmental monitoring	108
5.8	Environmental impact monitoring indicated by the district government	110
5.9	Constraints in environmental impact monitoring	111
6.1	Technical team perceptions on EIA stages conducted and stakeholders involved	123
6.2	Technical team perceptions on final responsible key parties for activities of the EIA process	124
6.3	Participation of stakeholders in the 4 coastal EIS review meetings	126
6.4	Inputs from stakeholders to the 4 coastal EIS review meetings	128
8.1	Recommendations for the improved use of environmental information in the coastal EIA process	158
8.2	Recommendations for improvement of stakeholder participation in the coastal EIA process	162
8.3	Main activities according to land use designation in the Banten Bay area	167
8.4	Main stakeholders for harbour and jetty development in Banten Bay	167

LIST OF FIGURES

		Page
1.1	Environmental impact assessment for tropical coastal projects	12
1.2	The Banten Bay study area	14
1.3	Structure of the thesis	18
3.1	Stages of Indonesian EIA process (based on regulation no. PP 51/1993)	45
3.2	Permit approval procedure for industry development as in the Ministerial Decree of Industry and Trade no. 256/1997	51
3.3	Implementation of the permit approval procedure as depicted in figure 3.2 for the establishment of a heavy equipment repair industry in the Banten Bay area	52
3.4	Implementation of the permit approval procedure as depicted in figure 3.2 for the establishment of a sugar refinery industry in the Banten Bay area	52
3.5	Permit approval procedure for jetty development as in the Ministerial Decree of Communications no. 58/1991	53
3.6	Implementation of the permit approval procedure as depicted in figure 3.5 for the construction of a jetty of the heavy equipment repair industry in the Banten Bay area	53
4.1	Potential effective use of coastal EIA to improve ICZM	61
4.2	The coastal EIA research in relation to the coastal EIA and ICZM	63
4.3	Studied elements of the use of environmental information in the research	64
4.4	Framework of the research	64
4.5	Assessment procedure for EIS review	69
4.6	Studied elements of stakeholder participation in the research	79
5.1	Average quality of the 10 coastal EISs reviewed differentiated for two sectors	91
5.2	Average quality of the 10 coastal EISs reviewed differentiated for two sectors and two dominant cases	93
5.3	Main sources of information used by consultant respondents to draft EIS	96
5.4	Main sources of information used by the technical team repondents to review EIS	99
5.5	Main sources of information used by the district government respondents to review EIS	102
6.1	Respondents' familiarity with the Jababeka project	131
6.2	Opinions of respondents toward the Jababeka project	131
6.3	Respondents' perception of their participation in the Jababeka project approval	131
6.4	Respondents' perception of their desire to participate in the Jababeka project approval	131
6.5	Stakeholder who made contacts with the district respondents regarding the Jababeka project	133
6.6	Perceptions of Serang district government on the importance of direct information sources	134
6.7	Perceptions of community respondents on the importance of direct information sources	135

6.8	Perceptions of respondents on the importance of written reports	136
8.1	Framework for improved use of environmental information in the coastal EIA process	154
8.2	Framework for improvement of stakeholder participation in the coastal EIA process	159
8.3	Framework for improved use of environmental information and stakeholder participation in the coastal EIA process	164

FREQUENTLY USED GLOSSARY

AMDAL: Acronym for *Analisa Mengenai Dampak Lingkungan* (see Environmental Impact Assessment).

Banten Bay coastal zone: a case study used in this thesis, located in the District of Serang, Banten province, Indonesia (reffered to Banten Bay area).

Bapedal: Acronym for *Badan Pengendalian Dampak Lingkungan* (EIA agency at national level).

Bapedalda: Acronym for *Badan Pengendalian Dampak Lingkungan Daerah* (EIA agency at district level).

Bappeda: Acronym for *Badan Perencanaan dan Pembangunan Daerah (*Planning Office at district level).

Bappenas: Acronym for *Badan Perencanaan Pembangunan Nasional* (National Development Planning Board).

BBIS: Acronym for Banten Bay Information System, a product of the Teluk Banten Research Programme, a research collaboration between Indonesia and the Netherlands (1997-2001).

BKPM: Acronym for *Badan Koordinasi Penanaman Modal* (Investment Co-ordination Board).

BPN: Acronym for *Badan Pertanahan Nasional* (National Land Registration Board).

Bulog: Acronym for *Badan Urusan Logistik* (National Logistic Board).

Competent authority: sector of government responsible for requiring EIA and reviewing EIS (based on government regulation no. PP 29/1986).

Coastal EIA: environmental impact assessment for development activities in the coastal area.

Coastal monitoring: monitoring of selected environmental parameters and indicators to determine quality of the coastal and marine components.

Coastal zone: the area at the interface between land and sea, where the sea influences the land and vice versa (Cicin-Sain and Knecht 1998).

Consultant: group of experts that on behalf of the proponent conducts the compilation of EIS.

Decentralisation process: the process of the handling of authority from the national government to the district government.

Depdagri: Acronym for *Departemen Dalam Negeri* (Ministry of Internal Affairs).

District area (*Kabupaten*): area under the jurisdiction of the district government.

District Government: local government at district level.

Effective EIA: EIA which promotes more sustainable and acceptable decisions among others supported by a good legislation and technical guidelines, a sufficient environmental information, adequate stakeholder participation, and a sufficient institutional capacity.

EIA: see Environmental Impact Assessment.

EIA Central Commission: a review committee developed at national level of the sector of government to review the EIS (based on decree no. PP 51/1993).

EIA District Commission: a review committee developed at district level to review the EIS (based on State Ministerial Decree of Environment decree no. 41/2000).

EIA guideline: a general guideline established to guide the EIA process, including the drafting of EIS, the reviewing of EIS, and the follow up of EIS approval.

EIA technical guideline: a sectoral EIA guideline translated from the EIA general guideline to implement the EIA process in one sector of activity.

EIS: see Environmental Impact Statement.

EIS review meeting: a phase in the EIA process after the completion of the EIS to determine the quality of the EIS as a basis for its approval.

EIS review package: a set of review criteria used to determine the quality of information presented in the EIS.

EMaP: Acronym for Environmental Management Procedures (*Rencana Kelola Lingkungan, RKL*).

EMoP: Acronym for Environmental Monitoring Procedures *(Rencana Pemantauan Lingkungan, RPL)*.

Environmental effect: a specific change in the environment caused by an existing or proposed development.

Environmental Impact Assessment (EIA): a process of assessing significant of impacts of a proposed development on the environment, which is required as part of the decision making process.

Environmental Impact Statement (EIS): a report as a product of the EIA process contains a detailed and in-depth research on the significant impacts of a proposed development plan.

Environmental information: information as listed in the EIA Indonesian general guideline that represented in environmental components; each component consists of several parameters, and each parameter consists of indicators.

Environmental information in the EIA guidelines: the level of detail of environmental information in the EIA guidelines specified by parameters and indicators.

Environmental information for environmental monitoring: the use of indicators of parameters for environmental monitoring and environmental impact monitoring.

Environmental monitoring: monitoring of selected environmental parameters and indicators to determine the quality of the environment.

Environmental impact monitoring: monitoring of selected environmental parameters and indicators to determine changes of the environmental quality caused by an existing or proposed development.

ICZM: see Integrated Coastal Zone Management.

Information sources (for EIA): sources of environmental information used by consultant to draft the EIS and used by technical team and district government to review the EIS.

Information sources to draft and to review the EIS: the use (use, need, availability and accessibility) of information sources to draft and to review EIS.

Integrated Coastal Zone Management: a process of governance for coastal and marine areas in order to optimise benefits from the coastal zone development and coastal resource management, and to minimise negative effects of such activities on the coastal resources and environment.

Institutional capacity in EIA: sufficient capability of district government on information management, on stakeholder participation, and to conduct the scoping and EIS review meetings.

Kabupaten: Indonesian term for District area.

Kepmen: Acronym for *Keputusan Menteri* (Ministerial Decree, or MD).

Marine information: environmental information related to the physical and biological components of the marine area.

MD: Acronym for Ministerial Decree.

MIS: Acronym for Management Information System.

MoI: Acronym for Ministry of Industry.

MoC: Acronym for Ministry of Communications.

Mitigation measures: measures to prevent and limit the negative impacts of the proposed development.

Monitoring measures: measures for regular data collection to assess the environmental changes caused by impacts of a proposed development.

Permit approval procedure: procedure for the approval of permits for a proposed development.

PP: Acronym for *Peraturan Pemerintah* (see Government Regulation).

Proponent: owner of the proposed development activity or plan subject to EIA.

Public access to information in EIA: rights of public to access information in the EIA process (based on decree no. 08/2000).

Public participation in EIA: participation of public to provide suggestion or comments on the conduct of the EIA and participation to be members of the EIS review committee (based on decree no. 08/2000).

Quality of the EIS: quality of environmental information as described in the EIS based on review criteria (Lee, 2000).

Regional Autonomy: Authority of the district governments to plan and manage their area (based on Act no. 22/1999).

RKL *: Rencana Kelola Lingkungan (*see EMaP).

RPL: *Rencana Pemantauan Lingkungan (*see EMoP).

SMoE: Acronym for State Ministry of Environment.

Scoping: the process prior to the impact assessment, by which to identify potential effects of a development via interconnections with other existing or proposed activities.

Scoring procedure: procedure to review the quality of the EIS involving scores.

Sectoral management: management of land and sea areas based on sector of activities, such as fishery, transportation, agriculture, etc.

Stakeholders: people and institutions who have a direct and indirect interest in the proposed development and its possible impacts.

Stakeholder participation in EIA: the active involvement and participation of the stakeholders in decision making in all stages of the EIA process.

Stakeholder participation in ICZM: process of involving stakeholders in decision making in ICZM process in order to foster sustainable development of a coastal zone.

Stakeholder participation measures: measures to improve stakeholder participation including measures to identify stakeholders, to raise participation awareness, to disseminate information and to invite input.

Technical team: a review team consists of EIA experts in various fields set up by the EIA commission to implement the review of the EIS.

SUMMARY

INTEGRATED COASTAL ZONE MANAGEMENT AND ENVIRONMENTAL IMPACT ASSESSMENT

Integrated coastal zone management (ICZM), is advocated by Chapter 17, Agenda 21 of UNCED held in Rio de Janeiro in 1992, to reduce conflicts of interest between coastal users on competition of uses for coastal and marine areas, adverse effects of one activity to another, and effects of activities to coastal and marine ecosystems. ICZM is the process that recognised the interactions between the natural and human systems, the balance between economy and development, the protection of environment and social structures, and interaction and consensus building between stakeholders. For more effective ICZM, its programs can be simplified among others through the use of environmental impact assessment (EIA) procedures. The EIA process for coastal development projects (coastal EIA) could then be used, as an efficient policy instrument to control coastal and marine management with emphasis on the assessment of impacts of coastal developments that are potentially adverse to the coastal and marine environments.

In order the coastal EIA to be used effectively to improve ICZM, it requires a good legislation basis, sufficient marine information, adequate stakeholder participation, and sufficient institutional capacity. Two important elements of the EIA that play important roles in the integration process in ICZM, environmental information and stakeholder participation are expected to perform significant roles in improving EIA effectiveness. This particularly relates to the four key areas of the EIA process: scoping, evaluation of significance, review of coastal Environmental Impact Statement (EIS), and monitoring and follow up of the coastal EIA decision.

ICZM AND EIA IN INDONESIA

ICZM has been advocated to reduce development pressures experienced in coastal countries like Indonesia. However, ICZM in Indonesia is not effectively implemented. Several reasons for this lapse have been identified of which issues on lack of communication and co-ordination among the government agencies are identified. Other institutional concerns such as conflicts in legislation, ineffective law enforcement, multi-sectoral and inter-sectoral integration and co-ordination, and conflicts have also been identified. An effective program for permit approval for coastal developments through the use of EIA would make the use of ICZM more effective.

Linking the Indonesian EIA with the approval system of project permits is considered vital for the enforcement of environmental standards and regulations. However, there are a number of obstacles to successful implementation, which would take considerable time and effort to overcome. In order to make the EIA more effective and useful to decision-makers, its procedures should be easily understood to facilitate actions on implementation of decisions. Also in Indonesia, environmental information and stakeholder participation are considered as important elements in the integration process of coastal EIA and ICZM, as well as performs significant roles in the coastal EIA to improve ICZM. This is also suggested by a pre-analysis in the Banten Bay coastal area. The limited participation of the district government in resolving the issues in the Banten Bay area is a problem to be managed. Addressing the present and future impacts due to development in the Banten Bay requires active interventions from district government - especially the District Planning Office (*Bappeda*), through the effective use of coastal EIA. The political and institutional changes of recent years such as the recent decentralisation of political power to the district level provide opportunities for active interventions. It is important for the district government to recognise the use of environmental information and the quality of stakeholder participation in the coastal EIA at district level.

CONTRIBUTION OF USE OF ENVIRONMENTAL INFORMATION AND STAKEHOLDER PARTICIPATION TO THE EFFECTIVENESS OF COASTAL EIA

Environmental information is essential to evaluate the impact of a coastal development activity and influences the evaluation of environmental impact management. It also improves decision-makers awareness on potential coastal environmental issues that are likely to occur in a proposed coastal development activity. Environmental information is an important input for monitoring and evaluation leading to modification or changes of the coastal development activity. In order to be effective, the coastal EIS should contain more information on the status of natural resources and environmental quality of the coastal area, with emphasis on marine biology components.

Stakeholder participation will improve the assurance that the significance of impact is properly assessed. When properly organised, stakeholder participation will also improve the flow of information among stakeholders helping proponents to better respond to their individual needs. For effective and efficient participation in the coastal EIA, the government and community should interact in order to enhance the decision-making process. All stakeholders interested in or affected by the coastal development issues should be included in the development process, continuously participate throughout the process, and also have access to information.

PROBLEM STATEMENTS AND HYPOTHESIS

The above analyses lead to the central problem statements of this thesis:

- How can environmental information and stakeholder participation contribute to a more effective use of coastal EIA for ICZM?
- How can improved environmental information and stakeholder participation be utilised by the district government for maximum benefit?

It is hypothesised that:

1. The use of marine environmental information in the coastal EIA process has the potential to improve the performance of coastal EIA and ICZM.

2. Stakeholder participation in the coastal EIA process has the potential to improve the performance of coastal EIA and ICZM.
3. The coastal EIA system has the potential to be utilised as a management tool to improve ICZM.
4. Environmental information generated by an integrated research team will contribute to better management decisions in ICZM.
5. The changes in policies within the new Indonesian government will have the potential to improve the role of district governments in coastal EIA and ICZM.

METHODOLOGIES OF THE RESEARCH

The general outline of the research is the analyses of the theory of EIA as laid down in a number of guidelines and the comparison with its practice. The research is restricted to two government sectors that play an important role in coastal developments in the Banten Bay, the Sea Communications and the Industry sectors. The assessments of the use of environmental information in the coastal EIA process consisted of 6 research activities:

1. Assessment of environmental information stated in 3 EIA guidelines,
2. Assessment of environmental information described in 10 coastal EISs,
3. Assessment of environmental information sources to draft and review EIS involving 59 respondents at the national and the district levels,
4. Assessment of the 4 coastal EIS review meetings at the national level,
5. Assessment of environmental monitoring involving 2 respondents at the district level, and
6. Assessment of perception of technical teams on the use of environmental information in the EIA process involving 4 respondents at the national level.

The assessment of stakeholder participation in the coastal EIA process consisted of 3 research activities:

1. Assessment of perception of technical teams on stakeholder participation in the EIA process involving 4 respondents at the national level,
2. Assessment of stakeholder participation in the 4 coastal EIS review meetings at the national level, and
3. Assessment of stakeholder participation illustrated by a case study involving 89 respondents at the district level.

The methods used consisted of interviews with open and close-ended questionnaires, observations, site visits, and publication reviews. The checklists included: checklists of parameters to review the EIA guidelines, a checklist of review criteria to assess quality of the coastal EIS, a checklist to observe the coastal EIS review meetings, and checklist of parameters to assess the environmental monitoring. This study is carried out under the umbrella of the KNAW/NWO/WOTRO Teluk Banten Research Programme, an Indonesian-Dutch research programme on Integrated Coastal Zone Management in the Banten Bay Coastal Zone (1997 - 2001).

USE OF ENVIRONMENTAL INFORMATION AND STAKEHOLDER PARTICIPATION IN THE COASTAL EIA PROCESS IN BANTEN BAY

- ***Limited availability of important information***
  The use of environmental information in most of the EIA stages was limited. The availability and use of the most important information for the coastal EIA such as environmental data, status of natural resources, and environmental monitoring data was limited. The EIA guidelines provide limited details on indicators of parameters within the biological component particularly marine biology, and also the activities referred to in the guidelines such as drafting and reviewing of reports and monitoring lack marine information. The district government has limited environmental data. The quality of the coastal EISs reviewed was not sufficient for a proper approval process. Basic information generated from the coastal EISs was not accessible and, even when available, often difficult to use as baseline data.

- ***Limited information sharing***
  The sharing of information between offices at local, provincial and central levels and even within the same office, which is important for the compilation and review of the coastal EIS and coastal management, was limited. The district government had limited access to local information on marine resources handled by the national and local sectoral offices or research institutions and the standard of data available in the research institutions did not suit the needs of the district government. Lack of methodology to exchange and regulate information sharing also worsened the opportunity of the district government to access the coastal information. This also resulted in the inability of the central government to access information available in the other institutions. Furthermore, lack of communication between the consultants of the feasibility study and of the EIA study was also identified.

- ***Deficiencies of the coastal EIA that constraint the use of environmental information***
  The opportunity for revisions in the coastal EIS review process was limited because the district government review of the EIS was done after it had been submitted to the central EIA commission. Standard review criteria that have been specified reduced the opportunity to improve quality of the coastal EIS. In some cases, the impact evaluation methods used in the EIS were limited and mostly related to qualitative rather than quantitative analysis. In some cases, data in some reports remained unanalysed. Evaluation of key impacts of the plan was not properly analysed in some coastal EIS and were limited to the normative method. The district government experienced limitations in coastal monitoring.

- ***Poor information management***
  Information management was poor due to reasons given below:

- The availability of local information on environmental data, environmental monitoring data and the status of coastal natural resources were limited.
- Sharing of information was lacking since information flow within and between offices was limited.
- Lack of clear policies and lack of resources within the district government was observed for example in the case of coastal monitoring.
- Fragmentation of responsibility and lack of integrating skills was indicated by the existence of conflicts in coastal and marine management over the land and sea areas, and over permits and EIS approval among national and local institutions.
- Lack of appropriate research and investigation in the study area was indicated by only one research programme: the Teluk Banten Research Programme being conducted within the area.

- ***The lack of empowerment of the participating stakeholder***

  The stakeholder participation in most of EIA stages was limited. The stakeholders was only expected to participate in providing data and information during the field survey while they had limited access to information concerning the development plan and the coastal EIA process. The participation of stakeholders in reviewing the coastal EIS was low and it did not influence the process of decision-making. Familiarity of the stakeholders with the project, their participation in coastal development approval and the communication process between the developer and the stakeholder concerning the plan were generally low. Awareness of the respondents on possible environmental impacts generated by the coastal projects as well as awareness on giving their input or information to the project was being ignored in the pre-coastal EIS compilation.

- ***Limitations faced by stakeholders***

  The results and findings of the coastal EIS were inaccessible to the stakeholders. The language, full of technical terms, used at the coastal EIS review meetings was another limitation on the district stakeholders to understand the EIS. For projects that are approved at national level or that are national funding projects, a national consultant was commonly hired. The distance between case locations limited the contacts or communications between the national consultant preparing the coastal EIS and district stakeholders. District government staffs, in particular, had not been sufficiently informed about the coastal EIA.

CONCLUSIONS

Based on thorough analysis of the research on the use of environmental information and stakeholder participation in the coastal EIA process, improvements of use of environmental information and stakeholder participation in the coastal EIA process for ICZM are essential. In the study several suggestions for improvement are given. A reflection is given to the central problem statement:

The use of marine environmental information and stakeholder participation has the potential to improve coastal EIA as an effective coastal management tool and to improve ICZM.

The research conducted for this thesis concluded that coastal EIA indeed can be used cost-effectively as a management tool to improve ICZM. Through improvement of use of environmental information and stakeholder participation in legislation and technical guidelines, institutional capacity, stakeholder participation, exchange and dissemination of coastal and marine information, and coastal monitoring, the district government will have an increased support for analysing and approving the EIS for coastal development, and thus the quality of the coastal EIA will be improved significantly. This will improve the quality of information on which coastal developments are approved and better quality of the decision on coastal zone development projects.

Environmental information generated by an integrated research team will contribute to better management decisions in ICZM.

In order to better understand a complex coastal system, the natural as well as human dimension should be studied in order to provide information needed by coastal managers. Banten Bay Information System (BBIS) is a prototype information system that aims to better understand the coastal processes and to assist the district government of the Banten area in conducting EIA and ICZM. The system contains information from the Teluk Banten Research Programme and developed in close collaboration with the district government. BBIS provides the district government with integrated environmental information on selected coastal and marine aspects on sea grass, fishery, coral reef, and birds as selected parameters for environmental monitoring. By making these results of integrated research accessible to coastal managers, BBIS contributes to the success of marine and coastal management conducted by the district government of the Banten area.

The changes of policies within the new Indonesian government have the potential to improve the role of district government in coastal EIA and ICZM.

The changes in Indonesian policies particularly in the decentralisation process in 1999 influence the implementation of acts and policies such as regulation and decrees related to the EIA. Decrees passed on the establishment of procedures to review EIS, on the formation of district EIA commission and on stakeholder participation in the EIA process have significantly improved the role of district government in coastal EIS approval. The establishment of Banten province in 2000, where the Banten Bay is located, significantly increased the power of district government to manage their marine and coastal area and directly to approve coastal development projects. This increased role for district government in ICZM increases the need for effective coastal EIA at the district level. The results of the study suggested that by having improved use of environmental information and improvement of stakeholder participation in the coastal EIA, the district government would have better support to use coastal EIA effectively to improve ICZM. The establishment of the Ministry of Marine and Fishery promotes the coastal and marine management and increases local autonomy and participatory of the district government in promoting ICZM.

PART I

INTRODUCTION

1. INTRODUCTION

1.1. Background

Integrated Coastal Zone Management

Coastal zones are the home of natural features such as beaches, estuaries, lagoons and coral reefs, as well as of human populations and their activities such as port development, commercial fisheries, industries, and tourism. The coastal zone, therefore, is ecologically and economically significant. Port development and dredging activity, for example, can significantly degrade coral reef through the occurrence of sedimentation. Reclamation activity can decrease seagrass and coral cover area through changes on current patterns and sediment transport. Industrial development in the coastal zone can generate pollutants including heavy metals and changes in water circulation and temperature patterns can change the productivity of wetlands. Similarly, the removal of mangrove forests for fish and shrimp ponds in tropical countries affects the functions of mangrove ecosystem. Activities in the further upland such as agriculture, industry, and mining for example, are sources of important pollutants that can affect the functioning of coastal ecosystem through the increased flow of sediments and pollutants into river and estuarine systems. These changes in marine and coastal ecosystem as marine life habitat cause depletion of coastal and marine resources. Increasing development activities and coastal resources management in the coastal zone, particularly in developing countries, have put the natural resources and marine ecosystems in these areas under severe pressure (Burbridge *et al.*, 1988; Cicin-Sain and Knecht, 1991).

The different coastal uses cause conflicts of interest between the users, including stakeholders, and government agencies at various levels. Conflicts of interest between the users, including direct and indirect users of particular coastal and marine areas involve competition of uses for coastal and marine areas, adverse effects of one activity to another, and effects of activities to coastal and marine ecosystems. Cicin-Sain (1994) argues that conflicts among government agencies at various levels occur due to various reasons, including different legal mandates and missions, different interest in the management of land-side of the coastal areas, different capacity on human resources, divergent in funding sources, and lack of information and communication. To reduce this pressure and conflicts, an integrated coastal zone management (ICZM), is advocated by Chapter 17, Agenda 21 of United Nations Conference on Environmental Development (UNCED) held in Rio de Janeiro in June 1992 and also by several international researchers afterward.

ICZM is the process that recognised the interactions between the natural and human systems, the balance between economic development, the protection of environment and social structures, and the interaction and consensus building between stakeholders It has been advocated to reduce pressure and conflicts of interest between the users, including stakeholders and government agencies at various levels over the different uses of coastal and marine areas. It has been promoted to develop public awareness, build capacity, and strengthen co-operation and collaboration between stakeholders, and integration of development and environmental protection. It requires the participation of all sectors and levels of government, integration of land and water boundaries, and an interdisciplinary and holistic approach in the decision-making process (White, 1989; OECD, 1993; Cicin-Sain, 1994; UNEP, 1995; Post and Lundin, 1996, Olsen and Hale, 1997).

The ICZM acknowledges the interrelationships that exist among coastal and ocean uses and the environments they potentially affect, and are designed to overcome the fragmentation inherent in the sectoral management approach (Cicin-Sain and Knecht, 1998). The ICZM incorporates a set of both substantive and procedural principles, a management strategy that emphasises adaptation and feedback, and the use of particular approaches, techniques and methods. It consists of the legal and institutional frameworks necessary to ensure that development and management plans for the coastal zone are integrated with environmental and social goals and the zone is developed with the participation of those affected (Post and Lundin, 1996). Objectives of ICZM are:

- to achieve sustainable development by taking into account the nature of the system, the balance between economic development and protection of the environment, and
- to reduce pressures of development on the coastal and marine areas.

The ICZM comprises horizontal, vertical and territorial integration (Cicin-Sain and Knecht, 1998). Horizontal integration harmonises different coastal and marine sectors, comprises integration of policies, development plans, investment, and management, as well as addresses conflicts among government agencies in different sectors. Vertical integration contains integration among different levels of government (national, provincial, local). It addresses differences of perspectives among multi-levels of government agencies in coastal and marine development policies and its implementation among different levels of government. Territorial integration or spatial integration consists of integration between the land and ocean sides of the coastal zone, integration among terrestrial, estuarine, littoral and offshore components.

ICZM initiatives are designed to develop public awareness, build capacity, and strengthen interagency co-operation and collaboration, and integration of development and environmental protection. In literature it is suggested that ICZM can be simplified into effective programs like Environmental Impact Assessment (EIA)-a procedure for permit approval in coastal development (Carpenter and Maragos, 1989; Maragos, 1989; OECD, 1993; World Bank, 1994; Clark, 1996; Frihy, 2001). Moreover, the use of EIA has been recognised by the World Summit on Sustainable Development held in Johannesburg in 2002, which strongly reaffirm the commitment to the Rio principles, the full implementation of Agenda 21 and the Programme for the further Implementation of Agenda 21 including on ICZM (World Summit, 2002).

Environmental Impact Assessment (EIA)

Environmental Impact Assessment (EIA) has been defined as a process of identifying, predicting, evaluating, and mitigating the biophysical, social, and other relevant effects of proposed development activities prior to major decisions and commitments being made. It is an environmental appraisal of the likely effects of a proposed development, and it has been internationally institutionalised and widely used as a formal procedure in development planning and resource management systems.

Mostert (1995) distinguished two objectives of the EIA:

- The immediate objective of the EIA is to facilitate sound and integrated decision making for proposed development in which environmental considerations are clearly stated.
- The secondary objective of the EIA is directed toward the achievement of environmental protection and sustainable development, among others, improving co-ordination among participating institutions, fostering better designed and cost-effective development projects, empowering community participation and building local capacity.

Sadler (1996) indicates a number of ways in which the EIA process is applied to decision-making based on her international review of environmental assessment system. The majority of countries use the EIA under formal institutional arrangements and which forms the basis for proposal and permit approvals, and the establishment of conditions for its implementation. These arrangements typically involve a framework of the laws, regulation, procedures, and guidelines. The aim is to follow a systematic procedure to ensure that specific proposals having potentially significant effects are identified subject to environmental impact assessment. The EIA provides decision-makers information on possible impacts of development plans so that correct decisions on the approval and implementation of these plans could be made.

It has been agreed upon by the UNEP teams developing "the Guide on Good EIA Practice" (Sadler, 1996) that the EIA process consists of three major stages:

- **Preliminary Assessment,** a process before the compilation and approval of the EIA reports which involves a screening to establish whether EIA is required, and scoping to identify the key issues and impacts that need to be addressed and prepare terms of reference for EIA.
- **Detailed assessment**, a process of the EIS compilation and approval that involves impact analysis to identify, predict and evaluate the potential significant impacts; mitigation to specify measures to prevent, minimise and mitigate the environmental damage; reporting or making of the EIS; EIS review to ensure the report fulfils terms of reference and standards; and decision making for EIS approval or rejection.
- **Follow-up,** a process after EIS approval that consists of monitoring to check that actions are in compliance with terms and conditions; management to address unanticipated impacts; and evaluation to EIS results.

The use of EIA for coastal projects or activities that are potentially harmful to the coastal and marine environments has been suggested in literature, and recently, it has been promoted by the Johannesburg Summit (World Summit, 2002). Since most countries have established EIA procedures embedded in permit approvals, the EIA process could be used as an efficient policy instrument to control coastal resource management. The coastal area supports a broad range of physical or a-biotic, ecological or biotic and social factors; therefore, it can be defined as an important part of the permitting system for the proposed coastal projects. The EIA for coastal and marine development activities (coastal EIA) is similar to the EIA for other areas, but it should concentrate on assessing the development impacts due to various activities conducted in the coastal area that potentially harmful to the coastal and marine environments (UNEP, 1988; Carpenter and Maragos, 1989; World Bank, 1994). The coastal area supports a broad range of physical or a-biotic, ecological or biotic and social factors; therefore, it can be defined as an important part of the permitting system for the proposed coastal projects.

Figure 1.1 shows an interaction process of a-biotic, biotic, and social aspects within the coastal EIA.

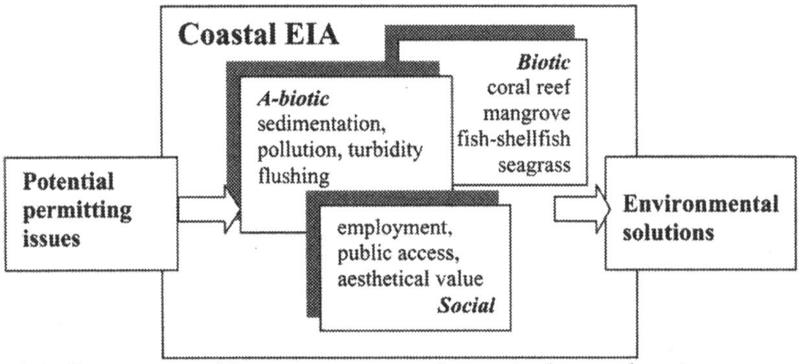

Figure 1.1. Environmental impact assessment for tropical coastal projects (After Maragos 1989).

The Indonesian EIA system

The Indonesian EIA (known as *AMDAL, Analisa Mengenai Dampak Lingkungan*) was introduced in 1982 with the passing of Act no. 4 on environmental management. It was implemented for the development of sector activities with the passing of regulation no. PP 29/1986 that sets out the general procedures for carrying out the *Amdal*, and regulation no. PP 51/1993 that gives authority for the *Amdal* process mainly to sectors of government at national level. All steps of the *Amdal* process from screening and scoping to approval processes are within the authority of the national level of government sector. There are many positive aspects in the design of *Amdal*, including the fact that it covers social and cultural as well as environmental impacts, provides inter-agency commissions to review individual project documents, and allows for public participation although not mandatory. Linking the *Amdal* process with the approval of project permits was considered vital for the enforcement of environmental standards and regulations. In the implementation, however, there are a number of obstacles that will take considerable time and effort to overcome as reported by World Bank (1993). The problems are often based on lack of information about environmental conditions and trends, especially those less visible to policy-makers and the public; uncertainty about ecological process and their reversibility, the potential costs and benefits of effective environmental protections; limited administration and financial resources for environmental protection within government institutions.

The environmental management based on sectoral approach as experienced by Indonesia hindered the integration of land and marine area management. The land and marine area are governed by different sector of institutions, in which each sector has its own procedure and guideline for the management, including the utilisation of EIA procedure. The management of coastal and marine areas is governed by the Act no. 24/1992 on spatial planning and Act no. 25/1992 on public participation which imply the separation of permit and the EIA for activities on land and sea areas and inclusion of public participation in planning and the EIA process. The Act no. 23/1997 on environmental management and regulation no. PP 27/1999 on the approval of the EIA reports clearly gives more power to the district government in the EIA process. The passing of the Act no. 22/1999 on regional autonomy gives more power to the district government level to manage land and marine areas with an integrated approach. For the EIA process, the act gives the responsibility of the EIA approval to the district government both for land and marine areas. Given the present responsibilities of district government, the capabilities of district government related to the EIA process for coastal development activities (coastal EIA) should be improved.

The Banten Bay study area

Banten Bay is located in the District of Serang, Banten province, Java, Indonesia (figure 1.2). It covers an area of 32,500 hectares and has a coastline of some 35 kilometres. The coastal zone is undergoing accelerated economic development with industrial estates being established along the western shores of the bay. This development is supported by infrastructure developments throughout the district and a major harbour in Bojonegara at the northwestern shore of the bay. Consequently, building of new residential areas and smaller industrial developments are going on in many places in the area. The spatial plans of the Serang District of 1994 and 2005 (Government of District of Serang, 1995a; 1995b)) show significant land use changes from mainly wetland agriculture to industrial areas (5.2%) and urban areas (9.3%). On the western shore, some 346 hectares of fishponds are designated for industrial use. New residential areas in the coastal zone are as large as 995 hectares. Figure 1.2 shows the Banten study area.

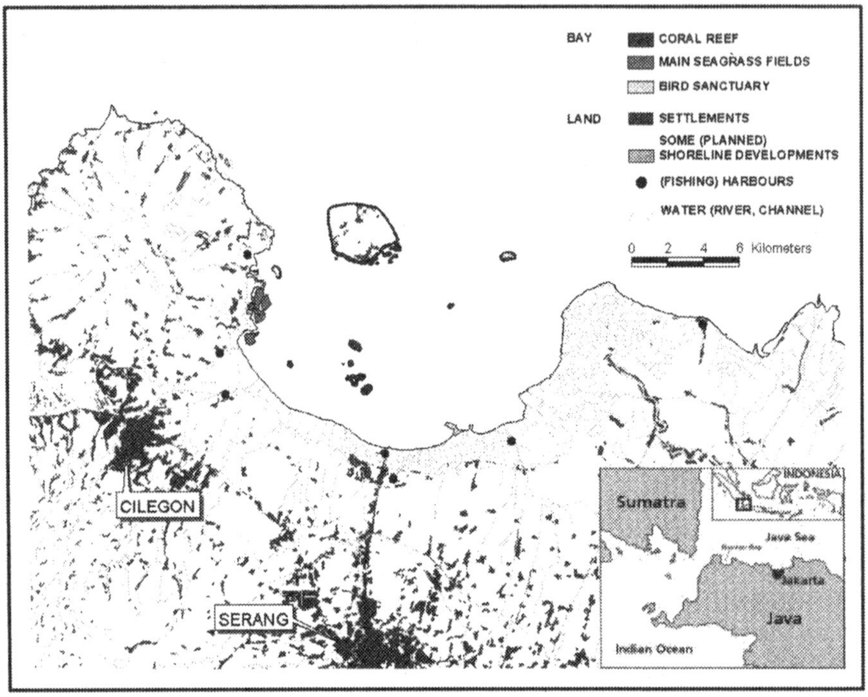

Figure 1.2. The Banten Bay study area.

Coastal development activities cause threats to the marine ecosystem. These range from land use changes due to expansion of industry, land reclamation and jetty development, to impacts of inland pollution and erosion (activities of industry, agriculture and domestic wastes) and over-exploitation of marine resources (Heun, 1996; Douven *et al.*, 2000). The growth within the area and future development have resulted in environmental conflicts in the form of community protests and the cancellation of a jetty development. There is evidence that urban and industrial encroachment into the coastal zone is leading to the pollution of estuarine and coastal waters. Pollution has affected flow and sediment discharges, which disturb marine resources (seagrass meadows, coral reefs, fisheries, the bird sanctuary of Pulau Dua and bird feeding areas) and increase their potential for exploitation. A significant area of seagrass field that functions as a nursery area for juvenile fishes of high economic value (Nuraini pers. comm., 1999) has faced a disturbance caused by past and present developments.

Teluk Banten Research Programme

The Teluk Banten Research Programme (1997-2001) was an integrated coastal zone management study conducted in Indonesia under research collaboration between Indonesia and the Netherlands by universities and research institutes of both countries. It was funded by the Foundation for the Advancement of Tropical Research (WOTRO) in the Netherlands and the Government of Indonesia. The Research Programme consisted of 9 research projects encompassing biotic, a-biotic, anthropogenic as well as more integral research projects. The 9 projects were: 3 a-biotic projects (sedimentation, morphological change, and water transport), 4 biotic projects (coral reefs, seagrass, groupers and snappers, and egrets), 1 remote sensing project, and 1 Integrated Coastal Zone Management project. The UNESCO-IHE, Institute for water education in Delft, the Netherlands, was responsible for the carrying out of the latter project.

One of the objectives of the research programme was to better understand the interactions between the a-biotic, biotic and anthropogenic systems. Another objective was to make this knowledge accessible to the district planning office of *Kabupaten,* Serang, which is responsible for the co-ordination of the coastal zone management of Banten Bay. This was done by means of the Banten Bay Information System based on studies of selected biotic, a-biotic and socio-economic processes and makes research results accessible in an ICZM and EIA framework. Several research components within the Teluk Banten Research Programme were placed in the integrative framework of the BBIS, including this EIA research.

1.2. Structure of the EIA research to support ICZM in the Banten Bay area

The primary objective of this research is to improve the EIA process and hence ICZM. To analyse the effective use of the EIA, this thesis focuses on the EIA process in the Banten Bay coastal zone. The objective of this research is reached by taking the following steps:

1. ***Elaboration of problem analysis on the basis of theoretical concept and pre-analysis of the Indonesian situation and the Banten Bay study area.***

In this step, theoretical concepts of ICZM and EIA will be introduced and an analysis of problems experienced by the Indonesian district government in the management of coastal areas will be given, using the Banten Bay area as a case study. These will be used to formulate central problem statements, hypotheses and methodologies.

2. ***Analysis of the practice of the EIA process in the Indonesian situation and in the Banten Bay study area.***

The analysis of the practice of EIA in this thesis will be based on case studies, which will focus on two sectors of government activities, the Sea Communications and the Industry sectors. The Indonesian EIA general situation and the Banten Bay study area will be used to test the proposed hypotheses. Two district offices will be used to focus the analysis of the research, namely, the District Planning office (*Bappeda*), which is responsible for planning co-ordination, and the District Environment office (*Dinas Lingkungan Hidup*), which is responsible for the implementation of the EIA process in the District of Serang.

3. ***Identification of recommendations and instruments for improved use of the EIA to achieve ICZM.***

The research results on the practice of the EIA process using the Banten Bay case study area, combined with the prospects of a decentralisation process, will be used to propose recommendations to the district and national governments. Instruments will be developed to support the recommendations.

1.3. Structure and Contents of the Ph.D. Thesis

This Ph.D. thesis consists of four parts:

I. Introduction,
II. Concept and pre-analysis,
III. Analysis of the case studies, and
IV. Conclusion.

An overview of the structure of this thesis is presented in figure 1.3.

The first part, the introduction of the thesis (chapter 1), introduces ICZM and EIA, the Banten Bay study area, and research on the coastal EIA process to support ICZM.

The second part, concepts and pre-analysis (chapter 2-4), describes the theoretical framework of the thesis. The analysis of management problems in the study area of Banten Bay is used to introduce the central problem statement of this thesis. It starts with the introduction of the concept on ICZM and is followed by the introduction of coastal EIA as an important tool to improve ICZM (chapter 2). The Indonesian situation is addressed in chapter 3. A preliminary research and analysis of management problems in the case study area of the Banten Bay (chapter 3) will be used to illustrate overall problems concerning ICZM. This discussion leads to the formulation of methodologies for the research (chapter 4) encompassing the central problem statements and the research design that will be used to test the hypothesis.

The third part (chapters 5-7) contains the analysis and discussions of the case studies. The analysis starts with assessments of the use of environmental information in the coastal EIA process (chapter 5). It consisted of 6 research activities:

1. Assessment of environmental information stated in the 3 EIA guidelines,
2. Assessment of environmental information described in the 10 coastal EIS,
3. Assessment of use of environmental information sources to draft and review EIS involving 59 respondents at national and district levels,
4. Observation of discussions in the 4 coastal EIS review meetings at national level,
5. Assessment of parameters for environmental monitoring involving 2 respondents at the district level, and
6. Assessments of perception of technical teams on the use of environmental information involving 4 respondents at national level.

The assessment of stakeholder participation in the coastal EIA process consisted of 3 research activities:

1. Assessments of perception of technical teams on stakeholder participation involving 4 respondents at national level,
2. Observation of stakeholder participation in the 4 coastal EIS review meetings at national level, and
3. Assessment of stakeholder participation in the pre-coastal EIS compilation, which will be illustrated by a case study involving 89 respondents at district level.

The discussion on the results of assessments of use of environmental information and stakeholder participation will be presented in chapter 7.

The last chapter (chapter 8) consists of recommendations for improved use of environmental information and stakeholder participation in the coastal EIA process and three instruments to support the implementation of selected recommendations.

The fourth part (chapter 9) is the conclusion. Solutions to the two central problem statements are presented. The concepts proposed and the relevance of their application are summarised in this chapter. The research design, the selected cases, and the methodology used in the research are discussed. Further research is suggested.

Figure 1.3. on the next page presents the structure of the thesis.

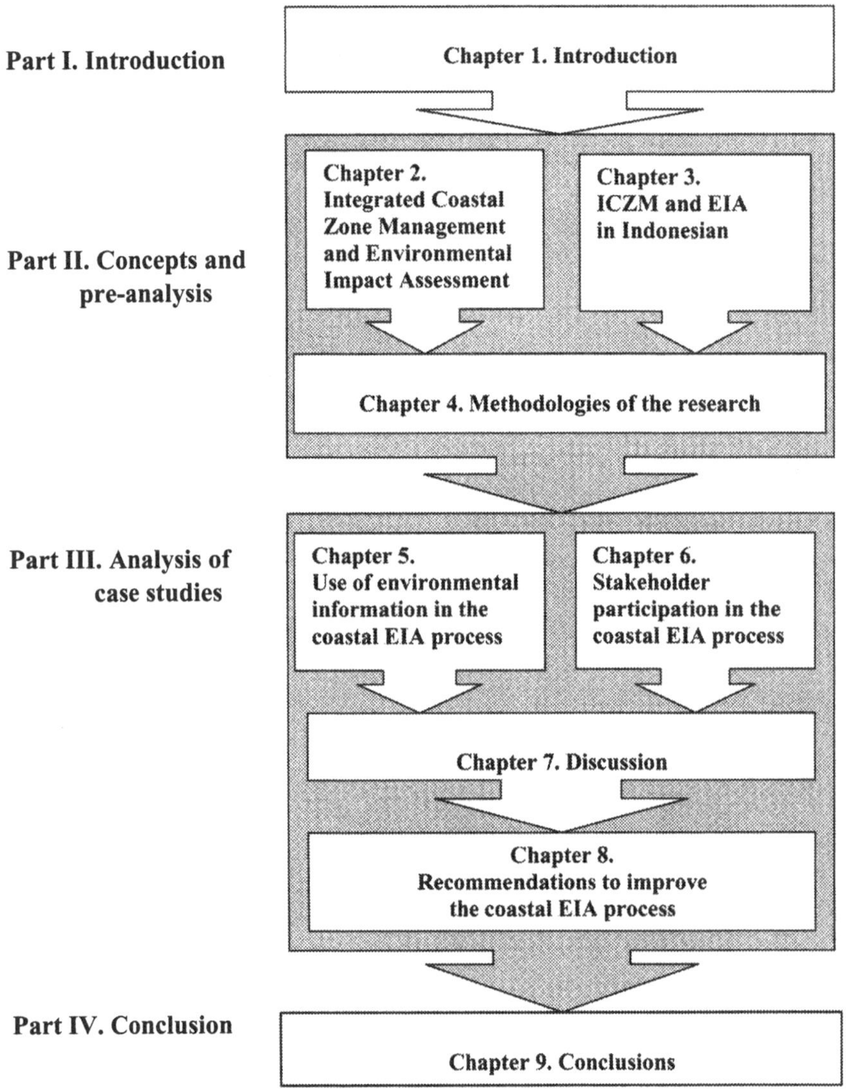

Figure 1.3. Structure of the thesis.

PART II

CONCEPTS AND PRE-ANALYSIS

2.
INTEGRATED COASTAL ZONE MANAGEMENT AND ENVIRONMENTAL IMPACT ASSESSMENT

2.1. Introduction

This chapter first presents a discussion of ICZM that covers decision-making in ICZM and important elements of the integration process in ICZM, involving environmental information and stakeholder participation (section 2.2). Then the linkage between ICZM and EIA is discussed, and a conceptual approach to the use of environmental information and stakeholder participation in EIA is presented (section 2.3). Finally an effective coastal EIA to improve ICZM is proposed for this thesis (section 2.4).

2.2. Integrated Coastal Zone Management (ICZM)

2.2.1. Decision-making in ICZM

ICZM is a continuous and dynamic process by which decisions are made for the sustainable use, development, and protection of coastal and marine resources (Cicin-Sain and Knecht, 1998). It should contain a continuous decision-making process involving integration of scales (local, regional, and national), themes (land and sea), stakeholders (corporations, households and government), and objectives (economic, environmental and safety). There are three elements of integration within ICZM, namely, policy, institutions, and planning as suggested by Vallejo (1993). Firstly, the national coastal policies should be integrated with the national ocean policy, in which it should be considered as a sub-set of a broader ocean policy that defines the role of the ocean and coastal areas as an integrated whole. Secondly, the institutional arrangement should include all levels of government, all sectors of government activities, and focus on two-way communication between various levels and sectors of government, both horizontal and vertical relationships. Thirdly, the planning efforts should include the formulation of a long-term coastal strategy as a basis for decision-making in the medium and short terms, and link the current and potential contributions of the coastal areas with the national development goals. Vallejo suggests that in order to assist in planning the integration and implementation effort, an information base is required that comes from many agencies having diverse responsibilities, and includes participation of various stakeholders involved in the management of coastal areas.

The decision-making process in ICZM involves multiple scales (local, regional and national), themes (land and sea) and stakeholders (corporations, households and government) having multiple objectives (economic, environmental and safety) (WCC, 1993; OECD, 1993; Cicin-Sain, 1994). Furthermore, the International Union for the

Conservation of Nature and Natural Resources (IUCN, 1990) identified two important features of decision-making in coastal management:

- the need for iterative feedback between components and the dynamics of interactions, and
- consensus building among all parties interested in the development of coastal policies and plans.

White (1989) mentions that the decision-making processes of ICZM link organisations of planning bodies, all sectors of government at various levels, private organisations such as companies, technical organisations (private institutes and universities), the community and NGOs. Similarly, Knecht (1994) suggests that the decision-making body should reflect the interests of the entire region, including the interests of the various components of the nation.

It has been agreed at various conferences that ICZM should focus on and represent a continuous and dynamic decision-making process (Cicin-Sain, 1993; OECD, 1993; Olsen and Hale, 1997). Decisions in the context of coastal zone management are normally not made by one stakeholder, but are the results of various actions and plans of responsible policy makers. ICZM can thus be regarded as a planning process with complex decision-making (Clark, 1996). Sorensen (1993) proposes two basic concepts of the decision-making process in ICZM:

1. Decision-making process requires the involvement of public and private bodies, and government and non-governmental organisations.
2. Decision-making bodies should decide and act in a co-ordinated, integrated way.

At the national level, ICZM should be understood as an effective means of addressing the problems resulting from fragmented sector-by-sector decision-making. At the local level, ICZM requires the decentralisation of government authority to local levels (Sorensen, 1993; Olsen and Hale, 1997). The role of the government in stimulating and supporting the integration process is prominent. This role involves different levels of government (Hofwegen and Jaspers, 1999) responsible for:

- The operational function (for example, carrying out services like education, irrigation and co-ordination of industrial activities).
- The organisational function (for example, the making of development plans).
- The constitutional function (providing legal and other frameworks).

The organisational function plays an important role in integrated coastal zone management.

As documented by Cicin-Sain and Knecht (1998), there is a great diversity of intergovernmental conflicts between developed with developing countries triggering the need for ICZM. Depletion of coastal and ocean resources, and the increase of coastal and marine pollution are typical conflicts experienced by developing countries. The typical conflicts in the developed countries were much more likely to be related to a competition to increase the economic benefits gained through the exploitation of the coastal and marine area. This is somewhat different to the common assumption that the concerns of developed countries would mostly be environmental pollution. Hale (2000) identified the presence of intergovernmental conflicts due to a number of reasons, among others:

- Jurisdiction for the management of coastal and marine area being within multi-levels of governments,
- Different objectives and priorities of national and local governments for the utilisation of coastal and marine resources, and
- Different capacity levels on expertise and information at varying levels of government.

The experience in the United States is that intergovernmental conflicts tend to be resolved by expanding policy and decision-making processes to include governments and organisations whose interests have been insufficiently accommodated in existing resource management schemes (Archer, 1989). Based on various studies, Bodani (2001) has identified some of the priority needs for integrated decision making in ICZM, namely:

- availability of sufficient information,
- adequate accessibility to local information,
- integration of information sources on social, economic, and environment,
- oriented research on problem solving,
- sharing of responsibility among multi levels of government, and
- strong policies on environmental management.

2.2.2. Environmental Information and Stakeholder Participation in ICZM

Environmental information and stakeholder participation are essential elements of the integration process in ICZM (Sorensen and Mc.Creary, 1990; OECD, 1993; Clark, 1996), in which use of the best information and the best science available as well as sufficient stakeholder participation is a fundamental belief in ICZM decision-making. Environmental information is needed to guide the wise use of coastal resources to protect the coastal environment and to improve the quality of life of coastal inhabitants. Stakeholder participation is needed to identify the information necessary for ICZM decision-making based on characteristics and perceptions of different stakeholders. Elaboration of environmental information and stakeholder participation is presented below.

Environmental information in ICZM

As discussed before, intergovernmental conflicts in ICZM among others are due to the different capacities for expertise and use of information at different levels of government. The environmental information refers to information on knowledge and factual data of the different environmental components related to development of the coastal zone. To lessen the conflicts, Cicin-Sain and Knecht (1998) suggest that an effective ICZM program must be based upon scientific information with respect to the physical, economical, social, ecological and governmental aspects of a coastal area. They argue that the scientific information is needed to guide the wise use of coastal resources, to protect the environment and to improve the quality life of the coastal community. Furthermore, a research activity can provide scientific information on matters related to the utilisation and management of coastal and marine resources in a given region. Based on research on the need of information in ICZM (Cicin-Sain and Knecht, 1998), the need of environmental information generated by research activities for ICZM is becoming evident as the complex relationship between economic development, environmental protection, sustainability of coastal resources, and welfare improvement of coastal population is recognised. Similar research on the need of scientific information in the ICZM has been assessed by the Joint Group of Experts on the Scientific Aspects of Marine Environmental Information (GESAMP). As stated in the report's executive summary, "scientific information is very important to understand ecosystem function and interaction process between human activities with the ecosystem, and for the resolution of coastal problems" (GESAMP, 1996).

Decision-making in ICZM should be based on sound research information, since it will support the solution of conflicts between various interests such as economics and conservation in coastal zone development as suggested by Cicin-Sain and Knecht (1998). Without adequate and accurate information and data for ICZM decision-making, controversial decisions to resolve conflicts cannot be implemented and enforced and domination of sectoral interests will be present in coastal development. Scientific information is also needed in the selection of management programs and in preparing materials for dissemination of information. The Thematic Study on Information (Doody *et al.*, 1999) has identified some of the strategic ways in which information assists the process of ICZM, among others:

- assists in improving awareness,
- helps to develop participation,
- provides the tools to identify policy and issue options,
- helps to provide a framework for decisions, and
- provides the tools for monitoring and evaluation.

Information generated from research concerned with the coastal and marine areas, such as coastal processes and construction activities, combined with information on the state of the economy, the marine environment and the effects of future developments will support ICZM. Post and Lundin (1996) indicate the kind of information that is required in the formulation of ICZM programs, including information on:

- coastal resource base,
- social organisations in the coastal zone,
- existing environment and resource-related programs, and
- institutional, legal and financial capacity.

Cicin-Sain and Knecht (1998), who based their research on a cross-national survey of ICZM, also identified types of data and information needed at various stages in ICZM, including information on:

- problems triggering the need for ICZM,
- resources, coastal uses, problems and conflicts,
- benefits and costs of ICZM and regulation and management measures, and
- monitoring of biological and physical parameters and social variables.

Elaboration of types of information as mentioned above will depend on characteristics and problems existing in each coastal area. Different types of information are needed in different stages of the ICZM for each coastal area.

The need of comprehensive information on ICZM is a complex issue because there are obstacles for using available information. Abundant volumes of information on coastal issues are available, however, the difficulties in dealing with coastal data is its complexity (Cicin-Sain and Knecht, 1998). The coastal data commonly consist of both spatial and temporal elements which represent complex relationships in the coastal zone (Clark, 1996). Furthermore, coastal information is commonly generated by a project research oriented project with focus on specific coastal components with limited consideration for future use (World Bank, 1993). Furthermore, Brown (1994) identified that limited information exchange between and within different levels of government has inhibited the use of available information. Another problem related to the use of environmental information in ICZM is not so much related to the provision and content of the information, but the way the information is presented to the decision-makers (Uhel, 1999). To deal with these problems requires knowledge on how the information is available at which location, and how it can be accessed.

In order to satisfy the need of information in ICZM, some important actions has been suggested by the report on the Thematic Study on Information (Doody *et al.*, 1999), among others:

- utilise the recent reports on coastal issue,
- create an information data base focus on main issue,
- build a regional common research program,
- ensure dissemination of research outputs,
- agree common protocols for information exchange, and
- set up networks on information exchange.

Stakeholder participation in ICZM

Stakeholders would be defined as those who involve directly and indirectly in activities dealing with coastal and marine area. They could include people and groups of people from all levels of governments, from conservation, industry and education communities (Bodani, 2001). Stakeholder participation relates to the process of involving stakeholders in identifying information needs, as well as the proper use of this information in ICZM. Co-operation among stakeholders is an important component of ICZM (WCC, 1993), which helps to avoid conflicts among the various interests in coastal management. In the context of ICZM, the characteristics and perceptions of the different stakeholders involved are an essential part of the information necessary for the integration process. Furthermore, Cicin-Sain and Knecht (1998) suggest that ICZM involves the integration of responsibilities of agencies at different levels of government, integration across different government sectors, local groups and sectors of the economy, as well as integration across technical, scientific and legal approaches. In particular, they propose that the district government is responsible for ensuring the participation of local stakeholders in ICZM.

The importance of the role of stakeholders in solving coastal conflicts has been recognised in various publications. A successful ICZM program is one that meets the needs of a large number of stakeholders, in which the decision-making processes effectively integrate both long and short term economic, environmental, social and equity considerations (Sorensen and Mc. Creary, 1990; OECD, 1993). Furthermore, Bodani (2001) suggests that the need to acknowledge the necessity of involving stakeholders in decisions and actions which affect them should be recognised, since broad participation of stakeholders in ICZM will provide an opportunity to resolve conflicts regarding the use of coastal resources. The stakeholders should be encouraged to participate in all stages of ICZM (Clark, 1996). Some ways and procedures for stakeholder participation are recognised in the literature. Stakeholders can participate:

- through consultations on resource management options,
- through direct involvement in the planning,
- through management and monitoring processes (Christie and White, 1997),
- through formation of committees to review the management strategies and their results,
- as advisory groups to the decision-makers (Bower, 1994), and
- as guardians in the monitoring process (Davies, 1987).

Some procedures for stakeholder participation in ICZM have been distinguished. Among these are:

- Enlarging the scope of involvement of stakeholders in the planning process,
- Providing an institutional framework and procedures to accommodate their participation (Stephenson *et al.*, 1994),
- Allowing all stakeholders to have access to relevant information (Davies, 1987),
- Establishing a chain of communication and procedures for decision-making,
- Providing mechanisms for integration lower and higher levels of decision-making, and
- Encouraging a participatory approach (Vallejo, 1993).

For other case study, Bodani (2001) has proposed several methods for empowering stakeholder participation in ICZM, among others through:

- The use of a round table model for developing a shared vision,
- The development of a co-operative partnership through consultation process,
- The use of accountability measures such as education program,
- The development of integrated management systems,
- The use of integrative strategies such as integrate information spheres and planning system,
- The stipulation of policy for stakeholder participation and for adopting good practice,
- The assurance obtained from stakeholders to contribute to baseline data needed for ICZM, and
- The identification of priorities for integrated decision-making.

2.3. ICZM and Environmental Impact Assessment (EIA)

2.3.1. ICZM and EIA

There are important components that should be taken into account in ICZM decision-making in order to achieve sustainable development, as suggested by researchers:

- the balance between economic development and protection of the environment and social values (World Bank, 1994; Clark, 1996; Zieren *et. al.*, 1998),
- the participation of all stakeholders (Stephenson *et al.*, 1994; Clark, 1996), and
- the incorporation of the EIA process in permit-approval procedures (Maragos, 1989; FAO, 1991; Frihy, 2001).

As discussed in section 1.1, international researchers have suggested that ICZM programs can be simplified into, among others, an effective EIA program (Carpenter and Maragos, 1989; Maragos, 1989; FAO, 1991; OECD, 1993; Clark, 1996; Frihy, 2001). They have suggested that ICZM programs can be simplified by including the following components:

- harmonisation of sector policies and goals,
- cross-sector mechanisms,
- a co-ordination office,
- permit approval, and
- EIA procedures in all decision-making processes.

The EIA is an anticipatory, participatory environmental management tool for reducing the negative impacts of proposed development activities, of which the EIA report is only one part. The objectives of EIA are to disclose to decision-makers and the public the significant environmental effects of the proposed activities, to foster interagency co-ordination and to enhance public participation (Mostert, 1995). Other objectives are to identify ways to avoid or reduce environmental damage by requiring implementation of feasible alternatives or mitigation measures, and to disclose to the public reasons for approval of projects with significant environmental effects (Biswas and Agarwala, 1992; Wood, 1995; Lee and George, 2000). The EIA is a plural process that is shaped by the interaction of many stakeholders involving different sector of activities. The perspectives of stakeholders on how well EIA works vary, reflecting different roles, background and experiences.

As a decision-making tool, the EIA has two main functions to decide upon the acceptability of a project based on: its environmental costs, and as a planning tool to minimise adverse impact cause by the said project (Werner, 1992). He summarised that in most of developing countries, the EIA was used a planning tool, because from thousands EIS reviewed during the last ten years, not a single report was rejected due to environmental reason. He suggests that the EIA should provide decision-makers information on possible impacts of development plans so that correct decisions on the approval and implementation of these plans could be made. In order to make the EIA useful to decision-makers, information generated in the EIA should be easily understood and facilitate actions on implementation of decisions. Two important elements of the integration process in ICZM (section 2.2), namely, environmental information and stakeholder participation will be the subjects of further discussion of EIA.

2.3.2. Environmental information in EIA

As discussed in the previous chapter, environmental information is the knowledge and factual data on the different environmental components related to the development of a proposed plan for all stages of the EIA process. Decision-making in the EIA process as in reviewing of the EIA reports (EIS), monitoring, and follow up of EIA decisions require the assessment of environmental aspects using environmental information. Wood (1995) stated that the decision-making within EIA process will involve a large number of trade-offs in the use of information: between simplification and the complexity of the reality; between the urgency of the decision and the need of further information; between facts and values; between forecasts and evaluation; and between certainty and uncertainty. The use of environmental information throughout the EIA process of screening, scoping, monitoring, and evaluation of EIA decisions produces the most direct effects that can improve the effectiveness of EIA as suggested by Mostert (1995). He wrote that utilizing of information on the relevant environmental facts of an activity can change the perceptions of different stakeholders on their ideas on the environment and can influence the evaluation of impacts generated by the development activity and its alternatives. The information use in the EIA process also can improve awareness of the decision-makers on potential environmental issues that are likely to occur. The information can promote or oppose the proposed development activity. He summarised that the information generated by EIA can change the evaluation of aspects of environmental impact management as well as can lead to modification or changes of the development activity.

Most researchers on EIA in developing countries agree on the main problems related to information which hinder effective use of EIA, among others are lack of information on environmental and social systems and lack of information sharing (Bisset, 1992). He summarised that lack of environmental information is an important limitation in developing countries, since very little is known on the important topics of structure and functioning of an ecosystem. Commonly, in developed countries, the nature of their environments is known. So they have sufficient baseline data, thus the compilation of the EIS is easier. In contrast, the limited comprehensive information available on the baseline data in developing countries is sometimes considered to be a big problem because it is expensive to obtain the information in terms of skills and resources. Another typical problem experienced in the developing countries is reluctance to exchange information (Bisset, 1992). He reported that this is the result of limited production of EIA reports that are not properly distributed. Thus it is still considered confidential. Similarly, Biswas (1992) has stated that a problem in EIA in developing countries related to information is limited availability of reliable data. He explains that despite the availability of data in certain offices the poor management and unnecessary classification as confidential make the data inaccessible. The reliability of the available data in developing countries are also in doubt due to data analysis being incorrect, improperly classified, or being collected in wrong places.

Similar to the environmental information need in ICZM, as discussed in section 2.1.2, environmental information is necessary for almost all of stages of the EIA process. This information exists in various formats such as policy and legislation, research reports, government reports, information systems. For coastal EIA, which is similar to the EIA for other areas, environmental information is used in the assessment of impacts of development in coastal and marine ecosystems (section 1.1). The information necessary for the coastal EIA is found in coastal EIA guidelines, in coastal EIS, in information sources used to draft and to review the EIS, in monitoring data, and in a follow up reports on coastal EIA decisions. It has been stipulated in several international agreements dealing with the protection of the marine environment that the coastal EIA guideline should address the assessment of effects of any activities likely to pollute coastal and marine environments (Ahmad, 1987). The coastal EIA guidelines, which will be used to guide EIA practice, to draft and to review the coastal EIS, and to monitor the coastal environment, should contain more parameters and indicators directed to address impacts of development in coastal and marine ecosystems than in land ecosystems. The coastal EIS produced will be used to guide the implementation of impact mitigation and impact monitoring measures for coastal development activity. Therefore, it should contain sufficient environmental information that can be used to guide the implementation of impact mitigation and monitoring measures. To improve the quality of the coastal EIS, information sources used to draft and review the EIS should have more information on the status of natural resources and environmental quality of the area in question that can be used as baseline information to assess the environmental impacts of the proposed development.

2.3.3. Stakeholder participation in EIA

Stakeholders of the EIA process are people and institutions who are concerned with the success of a project, both those positively and negatively affected by the project. Davies (1987) as well as Howlett and Nagu (1997) defined stakeholders as those people and institutions who have interest in the successful design, implementation and sustainability of a project. They include groups who have interest in the acceptance of the project proposal and groups responsible and concerned with the planning and implementation of the project. Groups interested in the acceptance of the project include proponents or initiators of the project, decision-makers at all levels of government, practitioners and consultants involved in conducting the EIA study. Groups interested in the implementation of the project include institutions and individuals responsible for planning and implementation of the project, and the community in the affected area that will be affected by project developments.

Stakeholder participation in the EIA process, in general, is increasingly recognised as an important element of the EIA. More and more countries have enacted national regulations and laws dealing with detailed procedures for public consultation and participation in EIA (UNEP, 1995). Some studies on the review of the EIA process for international funding agencies (Crooks and Foley, 1995; Bisset, 2000) show that projects involving public participation were more likely to achieve their objectives, and less likely to fail. ODA (1996) reports that, based on their experience, public participation has a strong potential to improve project design, environmental sustainability and social acceptability of the project plan. It helps to address the relevant issues, to ensure that the magnitude and significance of impacts have been properly assessed, and helps to improve the quality and acceptability of mitigation and monitoring measures. Public participation can also be utilised to improve information flows among stakeholders, and to help proponents to better respond to the needs of different stakeholders.

Stakeholder participation in EIA has been promoted through the establishment of guidelines. In practice however, the participation was consultative rather than participatory. Mwalyosi *et.al.* (1998) for example, based their review on over 30 EIA processes in Tanzania. They reported that stakeholder participation was included by only two out of 30 EIAs. The level of participation here was only consultative. 8 EIAs included some component of interaction process between the proponents and the local people, and the remainder included highly unsatisfactory level of stakeholder participation. Stakeholder participation in EIA can be hampered by many factors that may vary according to many circumstances. Hughes (1998) has identified several constraints commonly experienced by stakeholders in their participation, including time and money, language barriers and public presentation, education, cultural differences, physical remoteness, pressures imposed by the project cycle, mistrust and elitism, timing, ambiguity in legislation and guidelines, and poor presentation of EIA findings.

Bisset (2000) distinguishes 4 types of participation involving the public and NGOs in the EIA process, namely:

1. *Information Dissemination*: In this type of participation the degree of participation is more towards passive participation, since the flow of information is more a one-way process from the proponent or the agency to the public.
2. *Consultation*: Here a two-way flow of information between the public and the proponent or the agency takes place, in which the public has the opportunity to give comments on the proposal.

3. *Collaboration and Partnership*: This type involves the public in the EIA decision-making process, in the control of the EIA process and in joint responsibilities for the EIA results.
4. *Empowerment and Local Control*: This type involves the public in all of the EIA processes through public representatives.

Three stages of stakeholder participation in EIA have been distinguished by Soet (1987), based on her research in the implementation of the EIA system in the Netherlands:

- *Limited and controlled participation*: Participation usually takes place late in the process. The existence of conflicts and the quality of information being used for decision-making are poor. Decisions are taken unilaterally by the agency concerned.
- *Extended but still controlled participation*: Openness and a growing willingness on the part of the authorities to recognise the importance of public participation and a high information level, since it must comply with the standards of the regulations.
- *Integral Participation*: Stakeholder participation commences at the earliest phase and is continuous throughout the process. Decisions are made based on co-operation between stakeholders and the consensus of all the parties involved.

With regard to the relevance of participation to stakeholders in the EIA processes, various authors (for example: Davies, 1987; FAO, 1991; Clark, 1996) point out that participation should commence at the earliest stage and continue throughout the whole process. The government agencies can provide support and encourage efforts of the stakeholders to participate.

Successful effective stakeholder participation in decision-making of the EIA process requires the internalisation of basic principles of stakeholder participation (CIDI, 2000):

a. *Pro-activity*: government and civil society take initiatives to develop and enhance the decision-making process.
b. *Inclusiveness*: full participation of all stakeholders interested in or affected by the development issues.
c. *Shared responsibility*: the commitments, burdens and benefits of development should be shared equitably among government and civil society.
d. *Openness*: inclusive and continuous participation throughout the process in which the decision-making process is open to input and information.
e. *Access*: all stakeholders must have timely access, at the various levels of government, to information, to the political process, and to the justice system.
f. *Transparency*: productive relationships between civil society and government require that both be more accountable and transparent.

g. *Respect for public input*: for effective and efficient participation, there is assurance that contributions deriving from the implementation of measures for participation are evaluated, analysed and considered.

2.4. Effectiveness of EIA to Improve ICZM

2.4.1. Basic principles for effective EIA practice

Effective EIA has been frequently defined in literature. Mostert (1995) for example, suggests that effective EIA promotes better decision, in which decisions are taken more sustainable or lead to a more sustainable development, and more acceptable at least for the stakeholders directly affected, than decision taken without EIA. Furthermore, he proposed five preconditions that have to be fulfilled for EIA to be effective:

- awareness of environmental problems among the stakeholders dealing with EIA in particular the initiators and the competent authorities,
- supporting political culture is a positive value in which openness, co-operation, public participation is provided toward EIA process, and in which research and expertise are involved,
- well organised environmental movement in which public participation and expertise are aware of checking the quality of the EIA,
- efficient system of environmental management in which among others, institutions, participation, legal action, human resources, financial resources, and other environmental efforts are important in determining the decision making, and
- availability of sufficient environmental expertise and of data.

Based on input from practitioners over a series of study workshops on the effectiveness of EIA process, there are basic principles developed for effective EIA practice (Sadler, 1996).

- The EIA should be applied to all development activities likely to cause potentially significant adverse impacts; used as a primary tool for environmental management to ensure that impacts of development are minimised; and applied on the basis of well defined rules and responsibilities for stakeholders.
- The EIA should be undertaken throughout the project cycle with clear reference for project approval and impact management; utilised best practicable science and mitigation technology in correspondence with required procedures; and provide opportunity for stakeholder participation.
- Whenever necessary, it should address other related and relevant factors including social and health impacts; cumulative and long term large scale effects; with available technological alternatives; and consider sustainability of natural resources. Therefore, the EIS should result in accurate and appropriate

information presented in a clear, understandable and appropriate form for decision making; and resolution for disagreement among stakeholders.

- Finally, the EIS should provide the basis for environmentally sound decision-making; the design, planning, and construction of acceptable development activities; and provide appropriate follow-up requirements based on the significance of potential effects.

2.4.2. Problems hindering effective use of EIA

The effective use of EIA for achieving sustainable development particularly in developing countries has been hindered by several key problems as identified in literature. Bisset (1992) for example, has identified some problems, among others,

- lack of information on environmental and social system as baseline data,
- lack of trained human resources,
- limited awareness on EIA,
- lack of co-ordination among multi-levels government institutions,
- lack of willingness of proponents to internalise EIA requirements, and
- lack of information exchange.

Similarly, Sadler and Verheem (1996) identified main issues in the application of effective EIA in developing countries, among others,

- lack of mandate provided by law and regulations,
- lack of political will,
- the absence of reliable baseline data,
- lack of local participation,
- lack of awareness on EIA among government institutions,
- lack of environmental consideration on EIA decision-making; and
- very little EIA system monitoring.

2.4.3. Application of effective EIA

The effectiveness of the EIA application as suggested before requires four necessary components: appropriate timing in initial assessments; clear and specific directions in the form of guidelines; quality of information and product conformity with procedural guidelines; and good communication and accountability of the results of the EIA with the decision makers. The EIA can be used effectively if supported by, among others, satisfactory methods, effective procedures, supportive legislation, sufficient data, effective review process, effective co-ordination, auditing and monitoring, adequate public participation, and integrity of the EIA commission.

For coastal development activities, one objective of ICZM programs is to investigate the environmental impacts of proposed major development projects on coastal resource ecosystems, using the most appropriate method involving a permit system. In order that the EIA can be used effectively to improve ICZM, the EIA for coastal activities (coastal EIA) should produce a good EIS that provides sufficient information applicable to decision-makers and utilise sufficient stakeholder participation throughout the process. A good coastal EIS should contain a brief description of the proposed coastal development plan, prediction and assessment of possible impacts especially on the coastal and marine ecosystems, measures to prevent or reduce significant impacts, and a description of coastal monitoring plan. The information generated by the coastal EIA will provide the basis for the decision-making on whether the proposed coastal development plan will be approved, disapproved, or approved with some modifications of the plan or with improvements on impact mitigation measures.

2.4.4. Effective use of coastal EIA to improve ICZM

It has been proposed (UNEP, 1988; World Bank, 1993 and 1999; Gilpin, 1995) that the EIA can be used effectively if supported by, among others, satisfactory methods, effective procedures, supportive legislation, sufficient data, effective review process, effective co-ordination, auditing and monitoring, adequate public participation, and integrity of the EIA commission. Mostert (1995) furthermore says that EIA effectiveness can be achieved through the quality and use of information, use of EIA procedures, process of decision-making in drafting and reviewing EIS, improvement of activities such as monitoring and EIA legislation.

Based on the above discussions on environmental information and stakeholder participation, it is proposed in this thesis that for the coastal EIA to be used effectively to improve ICZM, it requires a good *legislation basis*, sufficient *marine information*, adequate *stakeholder participation*, and sufficient *institutional capacity*.

1. *Legislation basis*: A legislation basis is established through the clear support of state and national policies, as well as regional environmental plans, and the application of the EIA principles to all coastal policies, plans, programs and projects that are likely to have adverse effects on coastal environment.
2. *Marine information*: sufficient marine information is generated through the availability of sufficient environmental information on marine and coastal area and sufficient coastal monitoring activities.
3. *Stakeholder participation*: Adequate stakeholder participation is achieved through active public participation and integrity of the participants, the availability of avenues for public remedy and public inquiry, and the availability of all relevant coastal EIA documents to stakeholders.
4. *Institutional capacity*: Sufficient institutional capacity of the coastal EIA process, including technical capability of the review process, is achieved through the capability of district governments to manage information, improved stakeholder participation and competent multi-disciplinary teams to conduct scoping meetings and coastal EIS review meetings.

Two important elements of the EIA that play important roles in integration process in ICZM are environmental information and stakeholder participation. These two elements, the focus of this research are expected to perform significant roles in improving EIA effectiveness, particularly related to the four key areas of the EIA process:

- in the areas of scooping,
- in the evaluation of significance,
- in the review of EIA reports, and
- in monitoring and follow up of EIA decisions.

In the area of scoping and evaluation of significance, the environmental information is characterised in the use of EIA guidelines and information sources to make the EIS, while the stakeholder participation is represented in the project approval and in the making of the EIS. In the review of EIS, the environmental information is characterised by the use of information sources to review the EIS and the view of the technical teams, while the stakeholder participation is represented in the review of the EIS. In the area of monitoring and follow up of EIA decisions, the environmental information is performed in the environment and coastal monitoring activity by district government, while the stakeholder participation is represented in monitoring activity. The EIA effectiveness could be strengthened immediately and cost-effectively by the improvement of the use of environmental information and stakeholder participation in the area of scoping, evaluation of significance, review of EIA reports, and monitoring and follow up of EIA process (Sadler, 1996). The scoping refers to the open process for determining information necessary for decision-making, the important issues and concerns, the significant effects, factors, and alternatives to be considered, and the appropriate boundaries of an EIA study.

2.5. Conclusions

Based on the above discussions on environmental information and stakeholder participation, it is proposed in this thesis that for the coastal EIA to be used effectively to improve ICZM, it requires a good legislation basis, sufficient marine information, adequate stakeholder participation, and sufficient institutional capacity. Two important elements of the EIA that play important roles in integration process in ICZM, environmental information and stakeholder participation as the focus of this research are expected to perform significant roles in improving EIA effectiveness, particularly related to the four key areas of the EIA process: scooping, evaluation of significance, review of EIA reports, and monitoring and follow up of EIA decisions.

3.
ICZM AND EIA IN INDONESIA

3.1. Introduction

This chapter presents a description and analysis of management problems in Indonesia. First, Indonesian coastal planning system, laws and regulations and position and role of government will be discussed (section 3.2). Then coastal management problems in ICZM and in the Indonesian EIA system will be discussed (section 3.3). Furthermore, changes in policies and roles of district government will be discussed (section 3.4). Finally, in order to have more focus on problems, preliminary research in the Banten Bay area, involving the approval procedure and its implementation in two study cases, will be analysed (section 3.5).

3.2. Indonesian Coastal Planning and Management

3.2.1. Planning system, laws and regulations

The planning system of Indonesia is guided by the General Outlines of State Policy (*Garis-garis Besar Haluan Negara,* or *GBHN*) which determine a 5-year development-planning framework of objectives and goals. These goals are priorities for development, which are translated and implemented by sectoral government agencies into yearly programs and projects. Until the establishment of Ministry of Marine and Fishery in 1999, the management of coastal and marine areas is conducted by sectoral governmental agencies through their representatives at provincial and district levels. It is basically a top-down sectoral planning system. The land area is mostly governed by the Ministries of Industry and Trade, Agriculture, Communications, and Settlements and Infrastructure, as well as District Governments headed by *Bupati*, the Chairman of *Kabupaten* (District area). The sea area is governed mostly by the Directorate General of Sea Communications within the Ministry of Communications. The District Planning Office (*Bappeda*) co-ordinates the representative offices of the sectoral departments.

The development of coastal and marine regulations in Indonesia was influenced by the directives in the Program of Action for Sustainable Development (Chapter 17 of Agenda 21) set up by the United Nations Conference on Environment and Development (UNCED) in 1992. Most of the national laws and regulations established since then were aimed at the achievement of sustainable development in the form of environmental protection and resource conservation. The laws established range from general laws, as the Revision Act on Environmental Impact Assessment (1992), the Act on Protection of the Ecosystem (1994) and the Act on the Protection of Biodiversity (1995), to the more specific laws as the Ban on Catching of Certain Fish Species (1995).

Until recently, Indonesia has not yet enacted law on integrated coastal zone management. The law that integrates the land and sea areas in a spatial context is specified in Act no. 24/1992 on spatial use planning (*Pemanfaatan Ruang*). This act specifies the use of a comprehensive and integrated approach to the spatial management of air, land and water resources. At present the act is supported by the Governmental Regulation no PP. 69/1996 on public participation. The passing of the Spatial Use Planning Act, in addition to Act no. 25/1992 on public participation in planning, can be seen as a starting point for setting basic jurisdictions for the district governments' role in coastal and marine management. The government's role in coastal management is further enhanced through the passage of Acts and regulations on environmental management, as the Ministerial Decrees nos. 45, 46 and 47/MNLH/1996 concerning sustainable coastal management and the implementation of technical assistance for coastal programs (SMoE, 1996a). Government policies with respect to the environment are based on Act no. 4/1982 on Basic Provisions of Environmental Management (SMoE, 1996b). The most important policy based on this act is the provision of the Environmental Impact Analysis Study for development proposals based on Government Regulation no. PP 51/1993 concerning environmental impact assessment (EIA). The EIA process is included in the decision-making processes for the development approval procedure, to ensure that the potential impacts on chemical, bio-geophysical, socio-economic, and socio-cultural aspects are fully considered in the project planning. Further involvement of EIA in coastal management is through the passage of the Ministerial Decree of Environment no. 55/MNLH/1995 on procedures for a regional impact assessment and Ministerial Decree of Internal Affairs no. 98/1996 on the designation of local jurisdiction for environmental management. Act no. 23/1997 on environmental management provides the umbrella for integrated environmental management at national level. The passing of government regulation no. PP 27/1999 on environmental impact assessment and no. 25/2000 on government authority for environmental management support the implementation of the Act on environmental management. Act no. 22/1999 on regional autonomy gives power to district governments to manage their marine areas as far as 4 miles from the shoreline. The passages of laws related to coastal and marine management implies that district governments should be able to pool their coastal and upland resources in planning and management, and strengthen their abilities to accommodate, solve and prevent conflicts resulting from competition between land management and coastal management in their areas.

3.2.2. Position and role of government

Roles of the multi-level governmental agencies in the district coastal areas vary depending on the type of their activities. For industry development, the level of involvement of the sectoral government is based on the type of industry and amount of investment for that industry. The bigger the investment (and, often, the higher the pollution), the stronger the influence the higher level has on the role of the district level or the provincial level, or even the national level of the Ministry of Industry and Trade. The seawater area is mostly under the Directorate General of Sea Communications within the Ministry of Communications. The Forest Protection and

Nature Conservation Agency (*Perlindungan Hutan dan Pelestarian Alam,* or *PHPA*) within the Ministry of Forestry directs the conservation of coastal ecosystems.

District governments are important actors in the decision-making process of integrated coastal zone management. They have a better understanding of the problems and opportunities within the development program and the impact of government decisions on resource uses and the community. Various coastal management studies in Indonesia mention the importance of the role of district government bodies such as the District Planning Office (*Bappeda*) in the planning system (Chua, 1993; Kasim, 1993; Zieren *et al.*, 1998). The role of the District Planning Office (*Bappeda)* is basically based on Act no. 5/1974 on basic regional governance, Ministerial Decree no. 185/1980 on guidance for organisation and working procedure, and Government Regulation no. PP 6/1988 on vertical institutional co-ordination. According to District Regulation of Serang no. 3/1982, the roles of *Bappeda* include the co-ordination of planning activities of various local governments, research to support the planning activities, and the monitoring of the implementation of development planning at the local level. The role of the District Planning Office (*Bappeda)* in the planning process includes integration and co-ordination (Presidential decree of Indonesia no. 27 year 1980). Integration covers system, function and policy (further explanation is given in Chua, 1993; Kasim, 1993). Co-ordination includes vertical co-ordination from national to local level and horizontal co-ordination among various sectors of district government, to ensure consistency between policy and program implementation. Specifically in planning co-ordination, *Bappeda* is given the tasks of integrating and co-ordinating planning activities, developing land use plans and permit approvals, and providing environmental standards and guidelines as well as environmental information, which are intended to prevent or mitigate environmental impacts.

The involvement of various governmental agencies in coastal management is mostly in the form of establishing management strategies and instruments for spatial development and environmental protection. One management strategy is the EIA with its Standard Operation Procedures (SOP) format. The SOP format consists of Environmental Management Procedures (EMaP or *RKL*) and Environmental Monitoring Procedures (EMoP or *RPL*). The District Industry Office and the District Planning Office set the regulatory instruments for zoning and improved institutional linkages in the form of co-ordination activities as well as spatial planning (Taylor, 1993). The passage of Ministerial decree no. 55/1995 introduced new EIA procedures for integrated activities in industrial estates and multi-use activities, and in one regional area, it aims to include the accommodation of coastal development and to ease the EIA approval process. Act no. 23/1997 on environmental management specifies the EIA as part of the feasibility study and delegates authority for EIA approval to the district government. Its objective is to enhance the decision-making in the EIA process, as well as to enhance district government involvement.

3.3. Recent Changes in Indonesian Policies and Roles of District Government

3.3.1. Changes in Indonesian policies

Indonesian policies have experienced a significant move towards a more democratic system with the changes brought in by 3 presidential cabinets in the 3 years from 1998 to 2000. Significant new laws were passed during these years, including acts that aim to promote the decentralisation process in the governance system. Act no. 22/1999 on Regional Governance (known as Regional Autonomy) authorises a district government to manage its land area, including the sea area, as far as 4 miles from the shoreline. The formation of new ministry, the Ministry of Marine and Fisheries, and the closure of some sectoral ministries such as Public Works, Information, and Social Affairs in 1999 are also expected to support the decentralisation process. These changes in Indonesian policies will influence the implementation of acts or policies such as the one related to the coastal management and EIA system.

The changes implemented by the new cabinet when President Habibie succeeded President Suharto in May 1998 resulted in a significant improvement of the role of district government in the decision-making process. Act no. 22/1999 on Regional Autonomy overpasses the domination of sectoral governments at higher levels. The accompanying Act no. 24/1999 on Financial Sharing between national and provincial/ district governments will contribute to the financing of forthcoming activities. Earlier, a district was given a budget allocation based on its population, disregarding how much it contributed to the national GDP.

Further changes took place when President Wahid took over from President Habibie in October 1999 and these should speed up enhancement of the roles of district government. The formation of the Ministry of Regional Autonomy will accelerate the implementation of the Act on Regional Governance. The formation of the Ministry for Marine Affairs and Fishery will help district governments to manage the sea and coastal areas, which were under the domination of the Sea Communications department. The closure of sectoral ministries will decrease the number of district sectoral offices and avoid duplication of ministry offices (*Kantor Wilayah*) at the district level, which finally will smoothen the implementation of local government activities. The increase in the number of provinces from 36 at present to 50 in the coming years will help district governments to manage their areas thoroughly. The newly established province of Banten, where the Banten Bay is located, will significantly increase the power of district government in the District of Serang.

3.3.2. Influence of the institutional changes on the governance system and the EIA process

Concern has been expressed about some implications of the implementation of new acts in the year 2001, especially in regard to the Act on Regional Autonomy and the role of district government. Under this law, all public service delivery function except defence, foreign affairs, monetary, and trade policy, and legal systems will be decentralised to district level of government. The district governments, which in the past have been subjected to the guidance from the national government, will be responsible for most of development activities such as preparation of development projects including EIA approval. The implementation of Act on regional autonomy could give rise to conflicts between district governments over the marine areas that should come under their jurisdiction. Since the act states that the area 4 miles from the shoreline will be under the authority of one district government, there could be overlapping areas of jurisdiction with adjacent district areas. Also, sectoral governments may be reluctant to hand over revenues of coastal marine management such as revenues from harbour and coastal development, which are presently under the domination of the Department of Sea Communications, as well as revenues from mining activities that come under the Department of Mining. Since approval procedure of some development permits still stay within the national level, while the review and approval of the EIS is within the local level, it may resulted in a conflict between the national level on the implementation of permit approval with the district level on EIA approval. Dahuri and Dutton (2000) noted some interrelated and often overlapping issues on coastal management due to the establishment of several new laws. The identified problems are lack of knowledge of coastal and marine resources and process, ignorance of coastal and marine resources, lack empowerment of coastal communities and marine resource users, lack of institutional capacity, and lack of clarity regarding legal authority and planning frameworks for coastal management. Table 3.1 shows the changes in Indonesian policy since 1982 and their influence on the governance system and the EIA process.

Table 3.1. Changes in Indonesian policies since 1982 and their influence on the governance system and the EIA process.

Year	Change in policies	Influence on the governance system	Influence on the EIA process
1982	Act no. 4 on Environmental Management.	Approval is within the national level of sectoral governments, co-ordination for environmental management within the Ministry of Environment.	Development and implementation of policies on EIA.
1986	Government Regulation no. PP 29/1986.	Authority for EIA approval delegated to the Minister or to the Head of the Province, Departments and provinces are required to have EIA commissions.	EIA processes apply to all sectors of activity; EIA report is part of feasibility study. Permit is granted only after approval of EIA report.
1990	Revision of Government Regulation no. PP 29/1986.	No requirement for the responsible agency to demonstrate those public consultation procedures in the EIA processes has been followed.	Provisions for public notification, public access to information, and public involvement in the assessment.
1992	Act no. 24 on Spatial Planning.	Recommendation for EIA and permit approvals within the provincial/local levels, EIA approval within sectors of national/ provincial government.	Planning on land integrated with that on sea area implies the separation of permit and EIA for activities on land and sea areas.
	Act no. 25 on Public participation.	Public Participation in Planning and EIA processes.	
1993	Government Regulation no. PP 51/1993.	Integration of regional impact assessment into regional development planning set responsibilities of EIA to sectoral agencies.	Simplification of screening and assessment processes reduces time limits. New EIA procedures for multi-use, industrial estate, and regional projects.
1994	Ministerial Decree for Environment no. 13/MNLH/3/1994.	Membership of EIA commission at national and regional levels, role of local government in review and approval.	
	MD for Environment no. 15/MNLH/31/1994	Establishment of EIA commissions within sectoral ministries, regional government as an EIA member.	Framework for TOR, EIA, Environmental Management and Monitoring Plans.
1995	MD for Environment no 55/MNLH/11/1995.	No recognition of the importance of the regional EIA projects by the *Bapedal* (EIA agency) and District Planning Office offices.	General procedures for Regional EIA, Stages for Regional EIA are as that for EIA.
	-"- no 57/MNLH/11/1995.		General procedures for multi-sector EIA, various activities within one ecosystem.
1996	MD for Environment no. 39/MENLH/8/1996.	Public participation at EIA preparation and review stages.	Stipulates the types of activities subjected for EIA as part of screening stage.
	-"- no. 45/MNLH/11/1996.	Implementation by provincial and district governments.	Sustainable coastal management for tourism beach, harbour beach, and coral reef.
	-"- no. 46/MNLH/11/1996.	Formation of Steering Committee for sustainable coastal programs.	Implementation programs and technical assistance for sustainable coastal programs.

1996	MD for Internal Affairs no. 47/MNLH/11/1996.		The designation of provinces for coastal sustainable programs, West Java province.
	MD for Internal Affairs no. 98/1996.	District and provincial government involvement in: environmental enforcement, the provision of information, control and monitoring of implementation of EIA, enhanced public participation.	Guidelines for the formation of *Bapedal*/EIA agency at regional level, designation of local jurisdiction for environmental management.
	Chairman of *Bapedal* decree no. 299/1996.	Public participation has legal support in EIA preparation and review stages.	Technical guidance for the definition of social aspects of EIA.
1997	Act no. 23 on Environmental Management.	Delegation of authority for approval to local government Public participation and local government involvement.	Enhanced decision making in EIA process, resolution of any complaints and conflicts, and recognition of public participation in EIA
1998	Presidential Decree no. 196/1998.	Formation of *Bapedal* at provincial and district levels.	The establishment of *Bapedal* as an EIA agency.
1999	Act no. 22 on Regional Autonomy.	Single authority for the management of land and sea areas, EIA approval within the Head of District area, strong local participation.	No separation of planning and EIA for activities on land and sea area development.
	Act no. 25 on Financial Sharing.	Much more financing for District Governments in managing their land and marine areas.	
	Government Regulation no. PP 27/1999.	Access of information by the public. Public participation in all EIA stages.	Establishment of procedures to review EIA reports.
	Formation of Ministry of Marine and Fishery and closure of several ministries.	Decrease in EIA commission members. Strong power for District Government, more public participation. Policies related to marine activities is co-ordinated by Ministry of Marine and Fishery	
	Chairman of *Bapedal* decree no. 18/1999	Authority for District Office to review and approve EIA reports, monitoring the EIA implementations.	Working procedure for *Bapedalda* (District EIA office).
2000	MD for Environment no. 02/2000.	EIS review and approval process within district government	Guidance to review the EIS (Environmental Impact Statement) reports.
	MD for Environment no. 08/2000.	Public participation and access to information is obligatory in the EIA process.	Stakeholder participation and access to information in the EIA process.
	MD for Environment no. 41/2000.	Capacity building is needed by the district governments.	Guidance on the formation of EIA commission at the local level.
	Provision of Banten Province, November 4-2000.	Rearrangement of jurisdictions for district level, especially for outside Java Island, more power to the district level of government.	

Source: Documents review 2000.

3.4. Management Problems in Indonesian ICZM and EIA

3.4.1. Management problems in Indonesian ICZM

The general ICZM problems have been discussed in chapter 2. In Indonesian situation, as described by Kasim (1993), is that the various governments at district level are co-ordinated by the District Planning Office (*Bappeda*) through activities in regional planning, implementation, reporting, monitoring and supervision (District of Serang Regulation no. 3/1982). He mentions several shortcomings in the achievement of effective co-ordination: (1) different perceptions of the meaning and objective of co-ordination, (2) unclear job descriptions and procedures for program implementation within a sector of government, (3) resemblance and overlapping of programs and or policies, and (4) poor communication between governments. He concludes that the existence of various vertical and horizontal levels of government has resulted in a weakened implementation of co-ordination between and within the institutions, with the situation worsened by unclear distinctions of functions, mandates and responsibilities between institutions.

A study of the Straits of Malacca and Eastern Sumatra (Burbridge *et al.*, 1988) has identified the need for Indonesia to achieve integrated multi-sector involvement in coastal planning and management. The South East Asian research project in Segara Anakan in Cilacap (ASEAN/US-COMP/DGF, 1993) has identified lack of communication and co-ordination among the government agencies as the major limitation to coastal management, due to the overlapping of agency functions, authorities and jurisdictions. Other institutional concerns are recognised, such as conflicts in legislation, legal framework, ineffective law enforcement, multi-sectoral integration/co-ordination, poor institutional capabilities, low public involvement, and intersectoral conflicts. Zieren *et. al.* (1998) conclude, based on their Riau study, that other management problems like weak enforcement of spatial planning at district level and high inconsistency in government programs exist. Similarly, Dahuri and Dutton (2000) reported several issues concerning the management of coastal and marine area in Indonesia, including lack of participation and empowerment of coastal communities, lack of legal authority and frameworks on coastal management, lack of institutional capacity, and lack of co-ordination among sectoral government at various levels.

Table 3.2 provides a summary of management problems related to coastal zones.

Table 3.2. Summary of management problems related to coastal zones.

International (chapter 2)	**Indonesia**
Existence of multi-level sectors of government	Existence of multi-level sectors of government
	Mostly top-down planning approach
Separation of land and water areas in government authority	Separation of land and water areas in government authority
Insufficient information for decision makers	Insufficient information for decision makers
Insufficient integrated database	Insufficient integrated database
Mostly single management approach	Mostly single management approach
Complex approval procedures	Complex approval procedures
Competing coastal resource demands	Competing coastal resource demands
Limited public participation	Limited public participation
Limited integration and co-ordination	Limited integration and co-ordination
No link mechanisms for integration and co-ordination	No link mechanisms for integration and co-ordination
	Weak enforcement of spatial planning

Source: Literature Review 1999.

3.4.2. Management problems in the Indonesian EIA system

Based on literature reviews, the following discussion presents management problems experienced in the implementation of the Indonesian EIA system. As concluded in chapter 2, two out important elements within the EIA process (section 2.3) were used to focus the discussion on environmental information and stakeholder participation. Figure 3.1 presents stages of Indonesian EIA process based on regulation no. PP 51/1993.

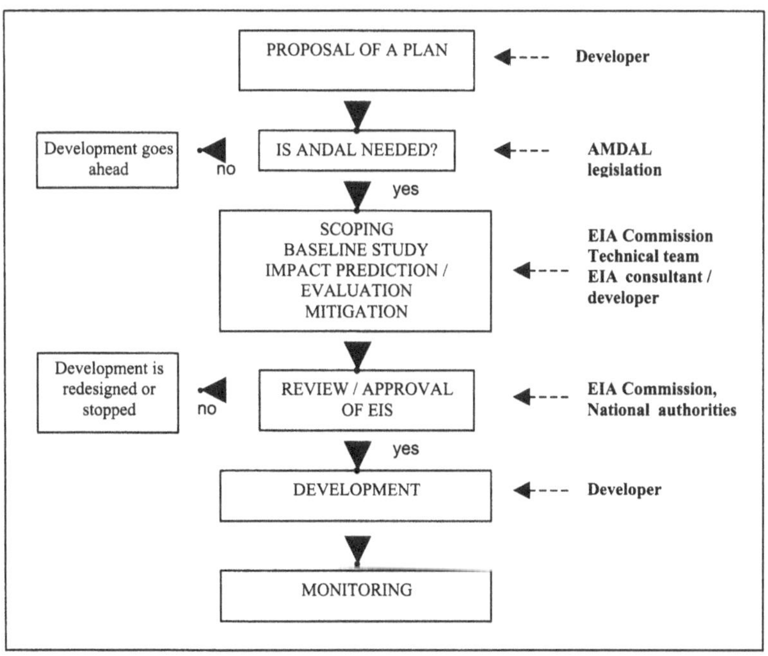

Figure 3.1 Stages of Indonesian EIA process (based on regulation no. PP 51/1993).

Problems in the use of environmental information in EIA process

Environmental information use in the EIA process is clearly stated in Government regulation no. PP 51/1993 on environmental impact assessment. Almost all of the EIA stages need the assessment of environmental information, except the screening stage that uses the available list of activities required from government sector offices. The EIS approval stage does not need environmental information since it is conducted based on recommendation of the results of the EIS review process.

Problems related to the implementation of EIA process in Indonesia are identified in the literature. International researchers have identified some shortcomings related to the use of environmental information in EIA experienced by developing countries, including Indonesia (Biswas and Agarwala, 1992; Wood, 1995; Donnelly *at al.*, 1998). The screening process, a process where the applicability of EIA is reviewed, is not conducted satisfactorily because a list of main criteria to decide whether or not the development plan requires EIA is not available. In relation to EIS preparation, baseline socio-economic and environmental data may be inaccurate, difficult to access or not sufficiently available. Historical data records needed for a proposed activity does not exist, and the EIA is conducted without such data (Achmad and Sammy, 1987; Tiwi, 2000b). The legal basis of the EIA systems may be weak, lacking in political backing, with little environment ministry influence and is often passed over by other more powerful ministries (Wood, 1995). Few EIA guidelines are implemented in practice because of their lack of relevance to the main tasks and problems facing guideline users, their general and non-specific nature, and lack of proper guidance in impact assessment (Donnelly *et al.*, 1998). The review of EIS appears to be facing problems like lack of expertise, lack of supporting data to crosscheck data presented in the EIS, and limited input from stakeholders (Achmad and Sammy, 1987; Bisset, 1992). Environmental monitoring is missing in the EIA process, the EIS reports are rarely available, and there is little interest in the monitoring due to limited resources and poor law enforcement (Wood, 1995; Tiwi, 1999b). Environmental information use in the EIA process is very poor. The offices responsible for EIA at national level are commonly working in conditions where environmental information is not accessible and information sharing is absent (Wood, 1995; SMoE, 1996b). The District Planning Office, *Bappeda*, faces constraints, especially in its role concerning environmental protection. Among these constraints are the lack of awareness for environmental protection and conservation, limited environmental information to support decision-making because limited studies are being conducted (SMoE, 1996b).

Table 3.3 presents EIA stages as defined by government regulation no. PP 51/1993 on the EIA process and need for environmental information.

Table 3.3. Stages of EIA process and need for environmental information (based on government regulation no. PP 51/1993).

Stages of EIA Process	Aspects	Need for environmental information
Pre-EIS	Assessment of aspects needed for plan development	Needed
EIS compilation		
1. Screening	Assessment of activities required EIA, activities generate significant environmental effects	Not needed
2. Scoping	Aspects identified: key issues, significant impacts, affected and interested population groups. Scoping of significant impacts: identification of potential impacts, evaluation of potential impacts, focus on significant impacts. Scoping of study area: project boundaries, ecological boundaries, social boundaries, and administrative boundaries.	Needed
3. Analysis/ prediction	Assessment of environmental/biophysical, social, economic, health, and risk. Effects and issues: direct effects, secondary/ indirect effects, cumulative effects, large scale/regional effects, biodiversity effects, and sustainability considerations.	Needed
4. Reporting	Prediction/estimation of environmental impacts/risks of proposal, probable effect on impacted public, prediction/ estimation of indirect and cumulative impacts, examination of alternatives, mitigation measures, evaluation/ determination of significant effects, recommendations, mitigation and monitoring.	Needed
5. Reviewing	Assessments of the quality of the EIS based on assessment of key impacts, recommendation for mitigation/monitoring measures based on the technical guidelines and environmental expert analysis.	Needed
6. EIS approval	Recommendation of the chairman of the EIA commission	Not needed
Post-EIS compilation		
1. Impact mitigation	Assessment of parameter and indicators for impact mitigation	Needed
2. Impact monitoring	Assessment of parameter and indicators for impact monitoring	Needed
3. Auditing and Evaluation	Evaluation of performances of impact mitigation and monitoring	Needed

Problems of stakeholder participation in EIA before the regional autonomy law

Indonesian regulations mandated public participation in the EIA starting in 1990, with the passing of regulation on public notification in the EIA process. The participation of district governments in the EIA review implicitly started in 1993 with the regional EIA procedure introduced by regulation no. PP 51/1993. The involvement of district governments in the EIA review process was explicitly stated by several government regulations in later years. Their involvement started with the appointment of district government as a member of the EIA commission in 1994, as an executor for sustainable coastal programs in 1996, and as a reviewer for the EIA reports in 1997. The most explicit involvement was the formation of *Bapedalda* (the EIA agency at district level) in 1999. Act no. 23/1998 on environmental management that is translated further by regulation no. PP 27/1999 on the procedure for EIA review explicitly addressed the requirement of public participation in all of the EIA stages.

Until recently, the concept of stakeholder participation was not commonly known at district-government level and it was not integrated into the different sectors of government policy. As a result there is limited knowledge and experience of non-governmental stakeholder participation. Public participation has been interpreted and implemented in various ways by various stakeholders, depending on the needs of each stakeholder. The degree of public participation varies; sometimes it is a one-way process with the public being informed of the development plans or the local stakeholders being briefed at EIA meetings. At national and provincial levels, the sectoral ministries dominate the involvement of government institutions in the EIA and in permit approvals (SMoE, 1996b). The acts and policies mentioned provide mechanisms for stakeholder participation. However, some issues still exist due to: lack of communication and co-ordination; conflicts in legislation and legal frameworks, multi-sector integration/co-ordination and institutional capabilities; low public involvement; inter-sector conflicts; poor enforcement of spatial planning; and lack of information. Various activities in nature conservation conducted by environmental non-government organisations, involving community participation (for example, in the designation of a nature reserve area) have proved effective in maintaining the sustainable use of natural resources.

Stakeholder participation in EIA after the regional autonomy law

The passing of the Act no. 22 on Regional Autonomy (1999) provided opportunities for public participation in the development of district areas as well as the EIA processes. The passing of several laws for the EIA process in 2000 increased opportunities for stakeholder participation in the EIA process. The degree of participation of stakeholders in the EIA process is defined by decree of Chairman of Bapedal no. 08/2000 on participation of the public in the EIA process and decree no. 41/2000 on the formation of the EIA review committee at city and district levels. Both decrees make it mandatory for district governments to review and approve the EIS. The decrees were not yet implemented when the research for this thesis was being carried out (1999-2000).

Stakeholder participation based on decree no. 08/2000 is limited to the participation of the public or affected community, in general. The public is defined as people concerned with the various effects of EIA decisions, due to their closeness to project location, on their economy, social practices, culture, environmental awareness, and other values or norms. The decree distinguishes between the public affected by the impacts and the public observer as well as defined. Two types of public participation in the EIA process are defined by the decree, namely:

- Participation of the public or community in the EIA process (informal participation) in which the public can pronounce their aspirations, needs and values, as well as their suggestions for conflict resolution.
- Participation of the public representatives in the EIS review process (formal participation) where public representatives are active in EIS review committee.

In this decree, the public is given rights to access information in the EIA process, to provide suggestions or comments on the proposed project, and to be members of the EIS review committee. Under the decree, the sector of government should inform the public on activities of the proposed plan and the EIS, compile and process all public inputs and present them to the EIS review committee, and facilitate public participation in the EIA process. The initiator of the project should inform the public on development activities before the EIS is composed, consult the public affected in the EIA study, and inform the public on the EIS and the proposed measures for impact mitigation and impact monitoring. Stakeholder participation based on decree no. 08/2000 is mostly considered as informal participation in which participation is not obligatory. The guideline is giving a suggestion to the initiator and to the sector of government to involve the public. The only formal participation is in reviewing the EIS in which the stakeholders are given the opportunity to review the EIS through their representatives. Table 3.4 presents stages of the Indonesian EIA process based on government regulation no. PP 51/1993 and stakeholder participation based on decree no. 08/2000.

Table 3.4. Stages of the EIA process (based on decree no. PP 51/1993) and stakeholder participation (based on decree no. 08/2000).

Stages of the EIA process (no. 51/1993)	Stakeholder participation (no. 08/2000)
Pre-EIS compilation	Informal participation
EIS compilation	
a. Screening	Informal participation
b. Scoping	Informal participation
c. Analysis/Prediction	Informal participation
d. Reporting	Informal participation
e. Reviewing	Formal participation
f. EIS approval	Not addressed
Post-EIS compilation	
a. Impact mitigation	Not addressed
b. Impact monitoring	Not addressed
c. Auditing/Evaluation	Not addressed

3.5. Preliminary Research in the Banten Bay Area

3.5.1. Study area

As presented in section 1.1, the Banten Bay area has been used as a case study area in this thesis. The area, with a coastline of some 35 kilometres, is located about 70 kilometres west of Jakarta. The bay has valuable ecosystems such as coral reef, seagrass area, fishing grounds and the bird sanctuary island, Pulau Dua. The study area is facing great pressure from present and future development activities due to, among others, environmental degradation caused by increasing activities of industries, development of a new harbour and land reclamation. The area is experiencing seawater pollution from urban and industrial activities that disturb marine resources.

3.5.2. Government strategies in coastal management

Strategies to manage the Banten Bay area were developed by the various sectors of district government. The District Industry Office is responsible for industry development includes granting Location Permits. The Public Works Office sets the policy for irrigation services. The Mining Office issues permits for sand mining on the land area, and the Directorate General of Sea Communications issues permits such as those for jetty and harbour development in the coastal area.

The District Planning Office of Serang has accommodated all the strategies to manage the Banten Bay including the coastal area in the Master Plan of the year 2005. The aim of the Master Plan is to provide guidelines for new developments to control spatial utilisation, and avoid duplication of new developments. The Master Plan gives priority to industrial development, which mainly be located in the western area of the Bay. A large number of Location Permits have been given for industrial development (5,050 hectares in *Kecamatan* Bojonegara-Ciwandan-Anyer and 11,932 hectares in *Kecamatan* Cikande-Kopo-Pamarayan). The development of large-scale industrial areas requires a change in the designated land use from irrigated agriculture to industry (for example, 400 hectares in *Kecamatan* Kramatwatu).

3.5.3. Development procedures in the Banten Bay area: two cases

Aims and methods

Two large industries located in the western part of the area, a sugar refinery and a heavy equipment repair industry, are analysed. These are considered typical for development within the area; in which the development involved land reclamation from the bay, jetty development and waste disposal that directly threatens a major seagrass field. The two cases are analysed to determine the management problems experienced, based on the official approval procedures and their implementation. The descriptions of the cases follow the structure of decision-making processes for successive stages of approval, as depicted in Figures 3.2 to 3.6 (after Mascia, 1997). The continuous cycle of the decision-making process, the participation of district government, and the use of EIA as well as information in the decision-making processes are considered. Information on these cases was gathered from interviews with district government officials, especially officials of the District Planning Office, and other related sectoral departments.

Development of the sugar refinery industry

The sugar refinery industry is using waste from the sugarcane industry as its main raw material. Liquid waste generated by the sugar refinery is very polluted. The sugar refinery is presently in operation, after all necessary permits were granted and an EIA report was approved. The industry experiences a conflict with the surrounding community because of suspected industrial pollution. The stages of decision-making for the approval of its development are: (1) Decision on granting a Principle Permit; (2) Decision on the Location Permit; (3) Decision on Regulation Permits; (4) Approval of the EIA; (5) Decision on an Operational Permit; and (6) Monitoring. This approval procedure is shown in figure 3.2 and its actual implementation is shown in figure 3.4.

Development of a heavy equipment repair industry with a jetty facility

This industry, which repairs all kinds of heavy equipment, needs a jetty to transport the heavy equipment. A 600-metre long jetty was built and some land reclamation was done. While the permit for jetty development has been granted, the establishment permits for the industry are being processed. The industry has a conflict with the surrounding community because both the new jetty and the land reclamation are suspected to have significant environmental impacts. On-going activities have now been stopped because no permit was given for the reclamation. The stages of decision-making for the approval of jetty development and industry development are: (1) Decision on granting a Principle Permit, (2) Decision on Recommendations, (3) Approval of the EIA, (4) Decision on Development Permit, and (5) Monitoring. The approval procedure for industry development is depicted in figure 3.2, while the implementation for a heavy equipment repair industry is in figure 3.3. Figure 3.5 shows the formal procedure for jetty development and figures 3.6 shows its actual implementation.

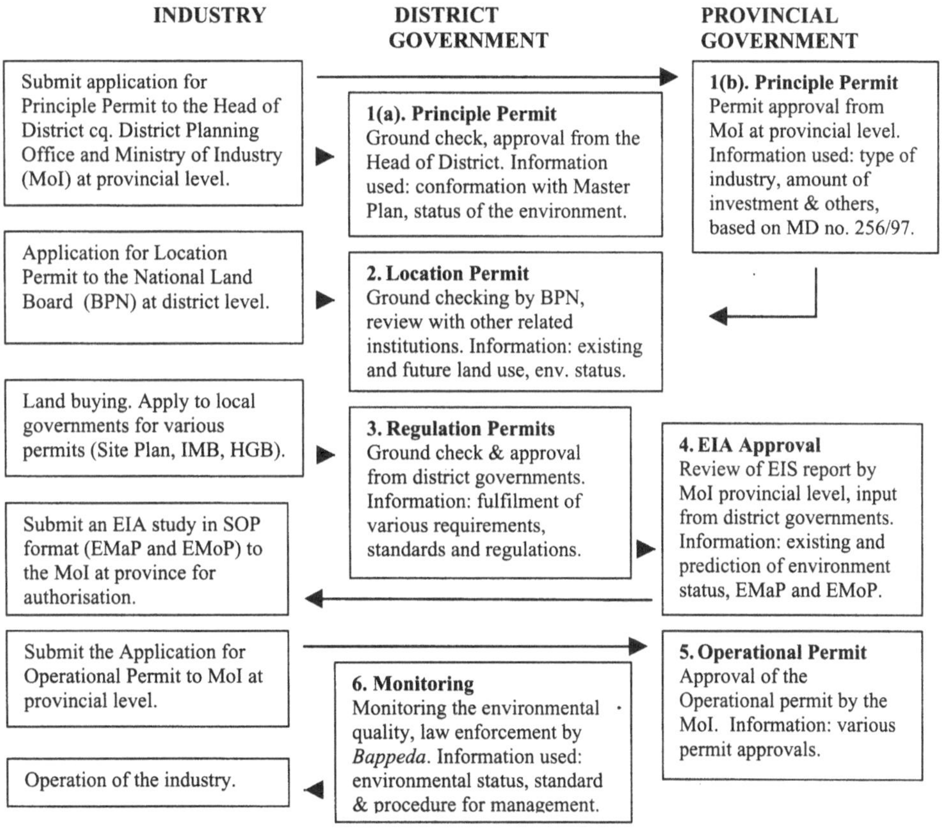

Figure 3.2. Permit approval procedure for industry development as in the Ministerial decree of Industry and Trade no. 256/1997.

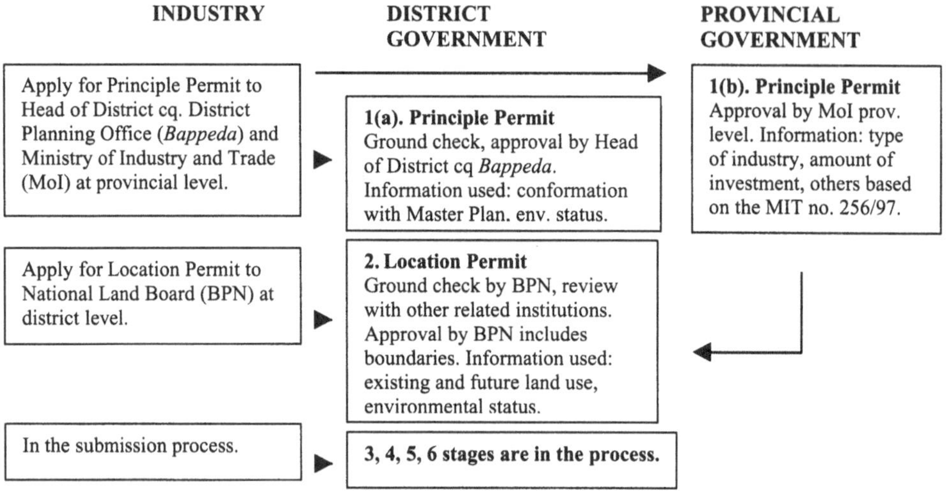

Figure 3.3. Implementation of the permit approval procedure as depicted in figure 3.2 for the establishment of a heavy equipment repair industry in Banten Bay.

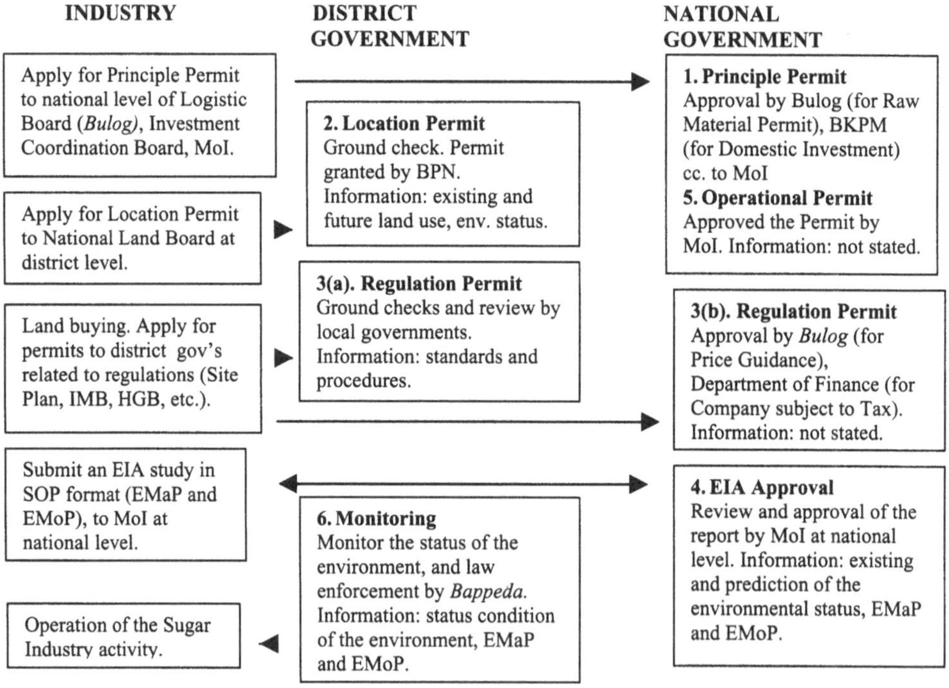

Figure 3.4. Implementation of the permit approval procedure as depicted in figure 3.2 for the establishment of a sugar refinery industry in the Banten Bay area.

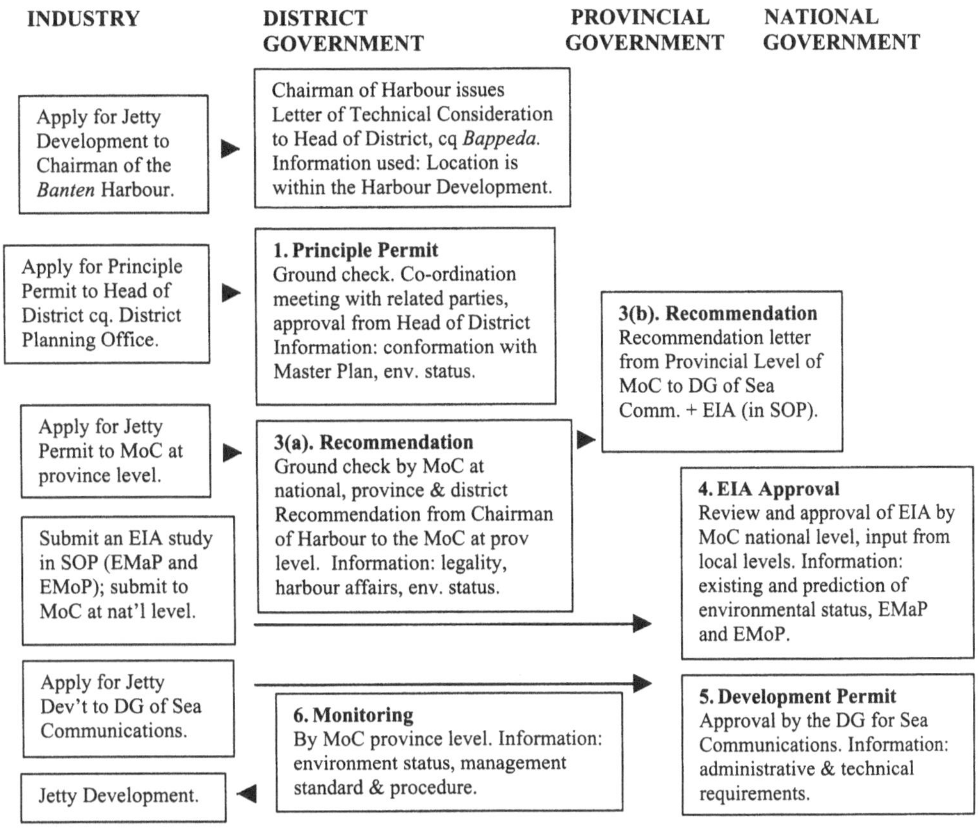

Figure 3.5. Permit approval procedure for jetty development as in the Ministerial Decree of Communications no. 58/1991.

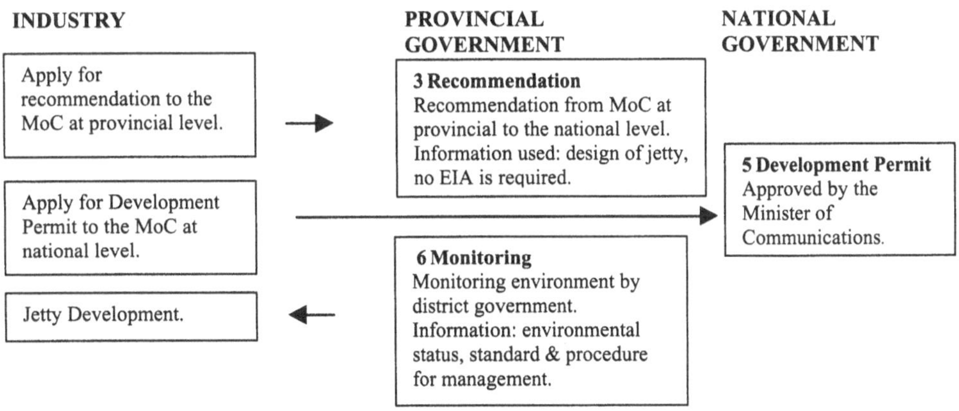

Figure 3.6. Implementation of the permit approval procedure as depicted in figure 3.5 for the construction of a jetty of the heavy equipment repair industry in Banten Bay.

Analysis and discussion of the two cases

As argued before, integrated coastal zone management requires the participation of all sectors and levels of government, the integration of land and water boundaries, and an interdisciplinary, holistic approach. The interactions between the natural and human systems, the balance between economic development, protection of environment and social structures, the related jurisdictions and the interaction and consensus building between stakeholders are all of concern to the decision-makers.

The two cases are analysed to determine whether there are shortcomings in the formal procedures and their implementation in the two cases. The discussion will concern mainly the decision-making process, the participation of district government, the use of EIA, and the information available on the components of the coastal system.

Permit approval procedure and in-sequence of its implementation

Figures 3.2 and 3.5 show that a cyclic decision-making process was present only slightly in Step 6, the review and approval of the EIA report. In this step, there was structural input from other sectors of government. For the other steps, it is mainly a one-way process from top to bottom within one sectoral agency. For jetty construction, there was also a bottom-to-top decision-making process within the Ministry of Communications.

The implementation approval procedure for the sugar refinery (Figure 3.4) did not follow the formal procedure. There was a tendency to bypass the formal procedures because Step 3, the Operational Permit, was granted at Step 1, which was supposed to be a Principle Permit only. It is not stated in the procedures what information is needed or what requirements have to be fulfilled for granting the permit, especially at the national level. Figure 3.4 shows that in the case of the heavy equipment repair industry, approval was carried out in accordance with the formal procedures. It should be noted that activities stopped because the industry went beyond its permits in land reclamation.

The approval for jetty construction as part of the heavy equipment repair industry did not follow formal procedures. The whole decision-making process was within one sectoral department. The mandatory recommendation from the Head of Banten Harbour and the local level of the Ministry of Communications, the Principle Permit from the Head of District, and the ground check by the district government are all missing.

Limited participation of the district government and community

Permit approval procedure, as in Figures 3.2 and 3.5, shows that most of the decision-making process takes place within one sectoral department. The Ministry of Industry at provincial level is the main decision-maker for the establishment of industries. The Ministry of Communications (MoC) is the main decision-maker for development in the coastal and marine areas. The District Planning Office *(Bappeda)* is represented as a consultant on whether the proposed location conforms to the Land Use Master Plan. *Bappeda* is only marginally involved in the granting of permits for jetty development.

Figure 3.4 shows that in the case of the sugar refinery industry, the main decision makers were at the national level, where 4 permits were granted at the ministerial level in the early stages of the permit approval process. There was no participation at the provincial level. The District Office of the National Land Board represented the only participation at district level. The District Planning Office *(Bappeda)* was involved only in the last stage, in granting the regulation permit concerning the Site Plan.

Figure 3.3 shows that in the case of the heavy equipment repair industry, the permit approval followed the formal procedures, which involve most of the district government offices, especially the District Planning Office.

Figure 3.6 shows that in the case of jetty construction, there was hardly any participation of district government as it is considered to take place in the coastal and marine areas under the jurisdiction of the Ministry of Communications within which all decisions are made. Since there is a separation of the procedures for land and coastal/sea areas, there is a tendency to misuse the permit, as was in the case for land reclamation. The permit was considered to be a location permit for jetty construction and not a location permit for the industry on the land site. Added to this is the weak monitoring of the national-level sectoral government that grants the permit.

The local community is not (directly) represented in the permit approval process. This lead to protests against the establishment of the industry and construction of the jetty. There was an incidental case of water pollution by the sugar refinery, which caused another protest. There was also a community protest against land reclamation by the heavy equipment industry, which was suspected to disturb the fishing grounds of the local fishermen as well as fill up the swamp area. As a result of these community protests, the local government discovered the violation of the permit and stopped the land reclamation activity. In this case, the involvement of the community was at the monitoring stage, the last stage of the decision-making process.

Lack of participation as experienced in the case study on coastal management occurred in other areas of Indonesia too as reported by Dahuri and Dutton (2000). They have noted some reasons for this as due to lack of empowerment of coastal communities and marine resource users which fact they attributed to the limited equity of coastal communities depending on coastal resource exploitation for livelihood despite the considerable efforts of government to address poverty of coastal community in the 1990s. Among the reasons for this situation are: limited of infrastructure, skill, technology, and capital experienced by coastal communities, as well as the dominance of national investor on coastal and marine management.

Limited use of environmental information

One of the attributes that characterises integrated coastal zone management is that the management strategies are evaluated on a system perspective, which recognises the interconnections among coastal systems (World Bank, 1996). Sorensen (1997) also recognised that countries with good practises in ICZM have utilised information systems and online data services, as well as information networks involving coastal policy and management.

Each permit approval procedure defines the information required specifically for that permit. However, this information is given in general terms, not in technical terms. For example, in the procedure for granting the Principle Permit by the Head of the District, the information provided only mentions the suitability with the Land Use Master Plan and no detailed information in technical terms is provided. The criteria based on which an EIA report should be approved or rejected are not formulated.

In the implementation of the two cases in Banten Bay, the details of information used for the same permit varied. The lower the level of government granting the permit, the more the details of environment information that was used. For the sugar refinery, the Principle Permit was granted at the national level and there is no stated information that was used for granting the permit. For the heavy equipment industry, more detailed information was used when the permit was granted at the provincial level.

The limited use of environmental information in coastal and marine resource management could be due to:

- lack of knowledge and awareness of coastal and marine resources, and
- lack of institutional capacity on coastal and marine management (Dahuri and Dutton, 2000).

This is despite the enormous efforts on capacity building conducted by several coastal research programmes (for example MREP, Proyek Pesisir), Dahuri and Dutton noted that a gap still exists in scientific skills and lack of experience on coastal management among the government at various levels. Furthermore, IRDA (2002) has noted that lack of co-operation and information sharing is some issues hampered the local government in the management of their area.

Limited co-ordination and integration in decision-making process

Most of the procedures describe a one-way decision-making process within one sectoral department. This reflects (1) the limited cyclic character of the process at different stages of decision-making and (2) the limited co-ordination and integration in decision-making between sectors of government.

There is a separation of responsibilities for planning and decision-making on permit approval procedures and for monitoring and management for the land area and water area activities. There is only limited participation of district government, vested in the District Planning Office, in permit approval and EIA authorisation. Most of the decision-making is within the higher level of government up to the national level. There is very limited participation of the local community and other local stakeholders or NGOs in the decision-making process. There is no mechanism for linking information in support of decision-making, because most of the permit approval is within one sectoral government department.

In implementation, there was a tendency to bypass the early stages of the procedures or to disregard the permit granted by the local level, especially, if the main decision-maker was at the national level. This is the case in industrial development. EIA reports were "approved" after the procedure on permit approval had been completed. The higher the level of the decision-maker, the more likely it was that supporting information was not considered. The higher-level sectoral departments, which granted the permits, did not have the capacity to monitor their implementation.

The limited co-ordination and integration in the decision making process as experienced in the case study can be attributed to as due to limited integration of initiatives on coastal management at various sectors and levels of the government. According to Dahuri and Dutton (2000) the traditionally sectoral nature of the government caused the limited integration of initiators in addition to project approaches experienced by the Indonesian coastal and marine management.

3.6. Conclusions

Based on the analysis and discussions on permit approval procedures and its implementation for the two cases as presented before, it can be concluded that implementation of an integrated approach using ICZM in Indonesia faces several constraints, such as limited participation of the district government and community, limited use of environmental information, limited co-ordination and integration in decision-making process, and domination of the national government in the issuance of permits and the approval of the development projects. Furthermore, there are two basic issues underlying the problems experienced on coastal management of the two case studies, namely: limited use of environmental information, and limited participation of district government. These two aspects are considered as increasingly important ingredients in the effective coastal EIA and the improved ICZM (chapter 1 and 2). They are the main factors hampering EIA and ICZM.

4.
METHODOLOGIES OF THE RESEARCH

4.1. Introduction

Following the discussion on ICZM and EIA and management problems in Indonesian ICZM and EIA in chapters 2 and 3, methodologies of the research will be presented in this chapter. Beginning with analysis of coastal issues and management problems (section 4.2) the central problem statements will be formulated (section 4.3), after which the research design involving the use of environmental information and stakeholder participation in the EIA process will be explained (section 4.4).

4.2. Problem Analysis

The analysis of coastal issues and management problems described in chapters 2 and 3 reveal that coastal management is a necessity and that it should be done in a framework of ICZM in order to be effective. An EIA is generally considered to be an effective tool to improve the performance of ICZM. Ideally, ICZM and EIA tools provide a framework for a planned approach to the development due to take place. Such a framework balances the socio-economic demands with the ecological capacity of the area. In the case of Indonesia, however, the practice of the ICZM approach can be considered as a new concept, which is not yet recognised as a useful approach by most government institutions and stakeholders. EIA, on the other hand, has a legislative status and is embedded in the permit system for economic development in Indonesia since the beginning of the eighties.

As concluded in chapter 3, implementation of an integrated approach using ICZM in Indonesia faces several constraints, such as:

- limited use of environmental information,
- insufficient participation of district government,
- limited co-ordination and integration in decision-making process, and
- domination of national government in the issuance of permits and the approval of development projects.

Meanwhile, the main problems hindering effective uses of EIA in developing countries are discussed in section 2.4.1, among others:

- lack of information on environmental and social system as baseline data,
- limited awareness on EIA,
- lack of co-ordination among multi-levels at government institutions,
- lack of mandate provided by law and regulations,
- lack of political will,

- the absence of reliable baseline data,
- lack of stakeholder participation,
- lack of awareness on EIA among government institutions,
- lack of environmental consideration on EIA decision making, and
- limited environmental monitoring.

One of the possible solutions to overcome the issues experienced in coastal management in the case study is the effective use of the EIA. The effectiveness of the EIA application as proposed by Mostert (1995) requires four necessary components:

- appropriate timing in initial assessments,
- clear and specific directions in the form of guidelines,
- quality of information and product conformity with procedural guidelines, and
- good communication and accountability of the results of the EIA with the decision makers.

As presented in chapter 2, Mostert (1995) suggests that the EIA can be used effectively if supported by:

- satisfactory methods,
- effective procedures,
- supportive legislation,
- sufficient data,
- effective review process,
- effective co-ordination, auditing and monitoring,
- adequate stakeholder participation, and
- integrity of the EIA commission.

As discussed in section 2.4, the effectiveness of the coastal EIA can be achieved through:

- the quality and use of information especially on coastal and marine area,
- the use of EIA procedures applicable for coastal area,
- the process of decision-making in drafting and reviewing coastal EIS,
- the improvement of coastal monitoring and EIA legislation, and
- the use of stakeholder participation throughout the process.

Therefore, in order to achieve effective results, this thesis proposes that improvement of effective use of coastal EIA should be conducted within the following areas:

- scoping,
- evaluation of significance of impacts in making the coastal EISs,
- review of the coastal EISs, and
- monitoring and follow up of coastal EIA decision.

As described in chapter 2, improvements in utilising environmental information and participation are essential elements of the coastal EIA effectiveness and of the process integrating coastal EIA and ICZM. Improved co-ordination of coastal development plans, permit issuance and environmental impact assessments for major coastal developments is necessary. The low participation of district government in solving the issues in the Banten Bay coastal area is a problem to be addressed, to ensure their participation in the coastal EIA process. In this respect, the district authorities should play a central role. At present, the national government dominates the increasing demands of coastal resources in the Banten Bay coastal zone area. Addressing the present and future impacts of development plans requires active interventions from district government, especially the District Planning Office (*Bappeda*). The political and institutional changes of recent years such as the recent decentralisation of political power to the district level provide opportunities for active interventions.

Figure 4.1 shows the potential effective use of coastal EIA to improve ICZM.

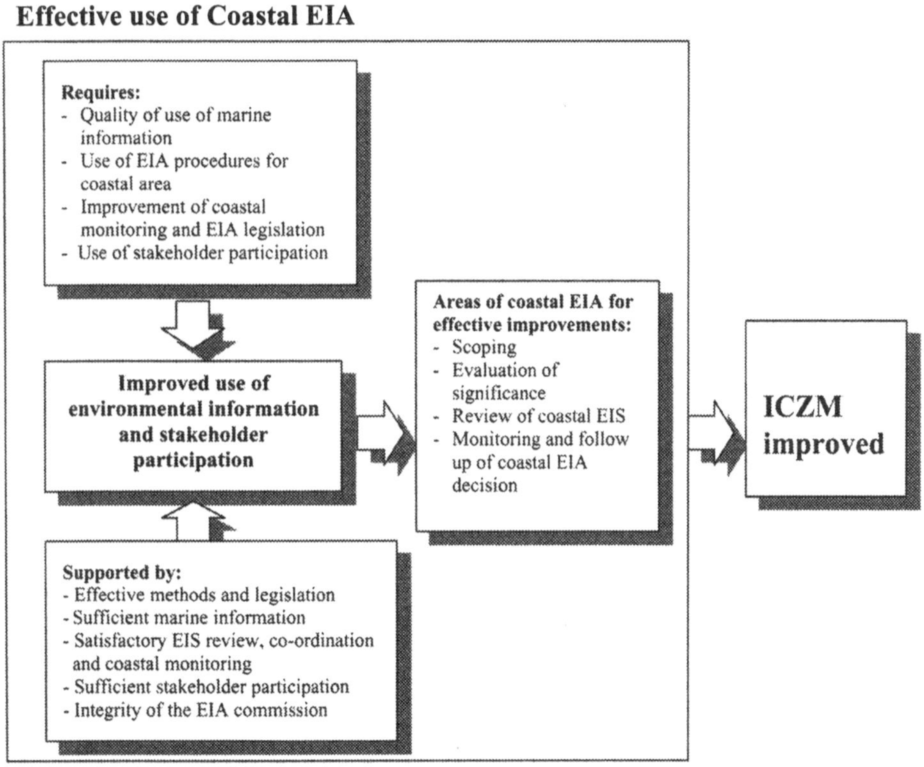

Figure 4.1. Potential effective use of Coastal EIA to improve ICZM.

4.3. Central Problem Statements

The analysis of management problems and the Indonesian EIA process, as discussed before, leads to the conclusion that improved environmental information and stakeholder participation in the coastal EIA process are essential. The coastal EIA process has potential to improve ICZM in which the district government can play an important role. Since the district government has been given the responsibility of the EIA process, it is important for the district government to recognise the quality of environmental information available with it and the quality of stakeholder participation in the EIA process in the district.

This leads to the central problem statements of this thesis:

- How can improved environmental information and stakeholder participation contribute to more effective use of the coastal EIA and ICZM?
- How can improved environmental information and stakeholder participation be utilised by district government for maximum benefit?

It is hypothesised that:

1. The use of marine environmental information in the coastal EIA process has the potential to improve the performance of coastal EIA and ICZM.
2. Stakeholder participation in the coastal EIA process has the potential to improve the performance of coastal EIA and ICZM.
3. The coastal EIA system has the potential to be utilised as a management tool to improve ICZM.
4. Environmental information generated by integrated research will contribute to better management decisions in ICZM.
5. The changes in policies within the new Indonesian government will have the potential to improve the role of district governments in coastal EIA and ICZM.

The general Indonesian situation will be used to test hypotheses nos. 1, 3, 5, and a part of no. 2 in order to have a more general overview. The Banten Bay area will be used as a case study to test hypothesis no. 4 and the remaining part of hypothesis no. 2. Figure 4.2 shows the coastal EIA research in relation to the coastal EIA and ICZM.

Sustainable development in coastal area

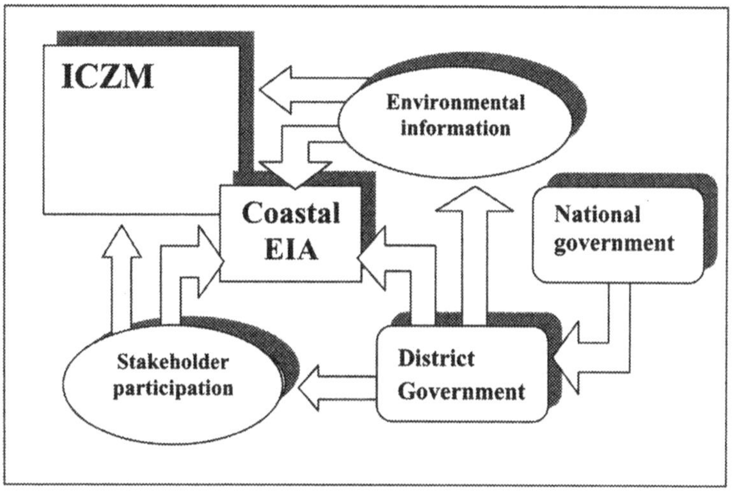

Figure 4.2. The coastal EIA research in relation to the coastal EIA and ICZM.

The general outline of the research is the analyses of the theory of EIA as laid down in a number of guidelines and the comparison with its practice. The research is restricted to two government sectors that play an important role in coastal developments in the Banten Bay, the Sea Communications and the Industry sectors. This study is carried out under the umbrella of the KNAW/NWO/WOTRO Teluk Banten Research Programme, an Indonesian-Dutch research programme on Integrated Coastal Zone Management in the Banten Bay Coastal Zone (1997 - 2001).

4.4. Research Design: Use of Environmental Information in the coastal EIA process

4.4.1. Objectives and research questions

Objectives

The quality of environmental information available to the district government is very important in the coastal EIA process. This brings us to the two objectives of this section:

1. to assess the quality of the Indonesian EIA in coastal zones, in particular, the use of environmental information, and
2. to provide suggestions for improving the use of environmental information in the coastal EIA.

Figure 4.3 shows studied elements within environmental information used in this research. They are based on analysis of case studies. Studied elements of the environmental information contained in the coastal EIA are the EIA technical guidelines, environmental information needs, availability and accessibility of environmental information.

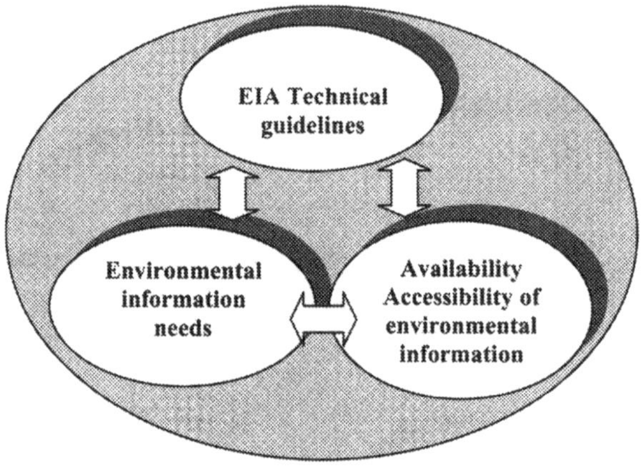

Figure 4.3. Studied elements of the use of environmental information in the research.

The use of environmental information in this research will be assessed based on information in EIA guidelines, in EISs, information sources used to draft and to review EIS, coastal EIS review meetings, environmental monitoring, and perception of technical teams. Figure 4.4 shows the overall framework of the research on the use of environmental information in the coastal EIA process.

EIA guidelines:		EIA practice:
• The Indonesian general guideline • The Sea Communications guideline • The Industry guideline	⟷	• Review of 10 coastal EISs • Sources of information used to draft and to review EIS (59 respondents) • Coastal EIS review meetings (4 cases) • Environmental monitoring (2 respondents) • Perceptions of technical teams (4 respondents)

Figure 4.4. Framework of the research.

Research questions

In order to explain the objectives of this research, the following questions are identified.

1. How is environmental information stated in the EIA guidelines?
2. What is the quality of environmental information described in the coastal EISs?
3. What environmental information sources are used by consultants to draft EIS and by the technical team and the district government to review it?
4. How is environmental information used in the coastal EIS review meetings?
5. What environmental parameters and indicators are monitored by the district government?
6. How is environmental information used in the EIA process based on perceptions of the technical team?

Methodological procedure
The first objective of this research is to assess the quality of environmental information in the Indonesian EIA process in the coastal zone based on the assessments of:

- EIA technical guidelines,
- coastal EISs as product of the coastal EIA,
- information sources to draft and to review EIS,
- coastal EIS review meetings,
- environmental monitoring, and
- perceptions of technical teams on the EIA process.

The EIA technical guidelines are one of several tools established and used by impact assessment practitioners to conduct the EIA study and by process managers to review the EIS. The EIA technical guidelines are assessed based on the level of detail of environmental information specified by parameters and indicators. An indicator can be measured, and a group of indicators will represent the quality of one parameter. A group of parameters will represent one sub-component of the environment.

The quality of the environmental information in the coastal EISs is identified based on the coastal EISs contents. The uses of information sources in the EIA are determined by consultants' use of information sources to draft EIS and also by the technical teams and the district governments to review EIS. The use of environmental information in the coastal EIS review meetings is determined by the discussions held. The environmental monitoring is assessed based on the use of indicators and parameters by district monitoring operators for environmental monitoring and environmental impact monitoring. The conclusions of assessment of results and also the perceptions of the technical teams are the basis for suggestions to improve the Indonesian EIA in the coastal zone.

Focus of the research
The research focuses on two sectors:

- the Sea Communications and Industry offices at national level, as the main sectors in the study area (see chapter 3), and
- the local governments in the District of Serang, as the case study area.

All research will focus on the practice of regulation no. PP 51/1993 on environmental impact assessment. The recent regulations no. PP 27/1999 on the review of the EIA reports, and ministerial decrees no. 40 and no. 41/2000 on the formation of the review committee at local level, were not implemented when the research was carried out (1999 2000).

Table 4.1 presents a summary of the methodologies used in the assessments of the use of environmental information in the EIA process. Elaboration of methods used in each research will be presented as follows.

Table 4.1. Summary of the methodologies to asses the use of environmental information in the coastal EIA process.

Research questions	Methods	Samples/Respondents
Objective 1:Assessment of use of environmental information		
1. Assessment of environmental information stated in the EIA guideline		
a. Assessment of the EIA guidelines	Checklist	3 EIA technical guidelines
b. Analysis of perceptions of technical teams on use of environmental information	Questionnaire	4 respondents at national level
2. Assessment of environmental information as described in the coastal EISs	Review package	10 EISs
3. Assessment of use of environmental information sources to draft and to review EIS	Questionnaire	59 respondents at national and district levels
4. Observation of discussions in the 4 coastal EIS review meetings	Checklist	4 meetings at national level
5.Assessment of environmental monitoring	Questionnaire	2 respondents at district level
Objective 2:Suggestion for coastal EIA improvement		
6. Assessments of perception of technical teams on the EIA process		
a. Conclusions of research results		
b. Assessment of perceptions of the technical teams on the EIA process	Questionnaire	4 respondents at national level

4.4.2. Assessment of EIA guidelines

Research question 1: How is environmental information stated in the EIA guidelines?

EIA guidelines are one of several tools established and used by impact assessment practitioners to conduct the EIA and to draft the EIS. Process managers also use them to review the EIS. The EIA guidelines set out the procedures and the mechanism through which information on environmental, social and economic effects of proposed activities could be presented clearly and systematically to decision-makers.

The aim of this research is to assess the level of detail of each parameter within the EIA guidelines, based on level of detail of indicators within that parameter. As the quality of the EIA guidelines influences the quality of the coastal EIS as well as the quality of environmental monitoring, the extent to which guidelines advocate the best practice for coastal EIS compilation and review needs careful evaluation. There should be an in-depth review of the level of detail of environmental information in the guidelines as specified in parameters and indicators.

Method of review

Checklists of parameters are used to review the EIA guidelines. This is done in order to simplify the review process, so that information generated can be easily structured and recorded. The checklists are used to review indicators of parameters as listed in the guidelines. The guidelines to be reviewed are the technical guidelines of the Sea Communications and Industry offices and the Indonesian EIA general guideline. The Sea Communications guideline is listed in the Ministerial Decree for Communications of the Republic of Indonesia no. KM 5/1996. The Industry guideline is in the Ministerial Decree for Trade and Industry of the Republic of Indonesia no. 250/1994.

The Indonesian EIA general guideline is in the State Ministerial Decree for the Environment no. KEP-30/MENLH/10/1999.

Development of checklists
Three checklists were developed for this research; each consists of a compilation of environmental components and parameters based on each guideline. All environmental components and parameters, as listed in the three review checklists, are adopted from the Sea Communications and the Industry technical guidelines and from the Indonesian EIA general guideline, with some adjustments. Each checklist consists of four review sections:

- parameters for environmental components,
- parameters for environmental impact components,
- parameters for impact mitigation and monitoring measures, and
- executing institutions for impact mitigation and monitoring, monitoring institutions, and related institutions.

Although the parameters for environmental components are the same as those for environmental impact components, the indicators within each parameter are different. For example, water quality is one parameter within the environmental and environmental impact components. Indicators for water quality of the environmental components consist of, among others, the levels of Biochemical Oxygen Demand (BOD) and Chemical Oxygen Demand (COD), while for the environmental impact components, indicators include the changes in levels of BOD and COD content.

Implementation of the method
Two EIA experts together with a researcher (*Note*: "researcher" refers to the author of this thesis) conducted the review of the guidelines in 3 steps:

- review of indicators of parameters from the two technical guidelines (conducted by each expert),
- summarising of results of the two reviews for each technical guideline (conducted by the researcher), and
- review of indicators within parameters of the Indonesian EIA general guidelines (carried out by the researcher).

The experts were asked to indicate the sufficiency of each parameter in the guidelines, based on the level of detail of environmental information specified in indicators. A parameter is indicated as sufficient if it has sufficient indicators to describe its quality. For those parameters that were indicated as insufficient, the experts were asked to make an addition of indicators.

4.4.3. Assessment of coastal EISs

Research question 2: What is the quality of environmental information described in the coastal EISs?

The Indonesian EIA system requires the submission of an EIS to the EIA central commission of the sector ministry for review. When this research was carried out in 1999-2000, there were no specific, written criteria to review the substance of the EIS. However, recently a Ministerial Decree for Environment (no. 2/2000) concerning guidance to review the EIS has been issued. The administrative regulation per sector often served as a guideline for reviewing. In the absence of review criteria, regulation no. PP 27/1999 on procedures to review EIS is generally concerned with procedural aspects rather than substance of the EIS. Since no standard scoring procedure is available to review EIS, there is a doubt about the quality of EIS and its utility value as a guide for impact mitigation and impact monitoring measures. The aim of this research is to review the environmental information, as described in the EISs, by in-depth analysis using an Environmental Appraisal Review Package.

Review package

The Environmental Appraisal Review Package used for this research contains a list of review topics and was adopted from the report of the EISs review study (Lee *et al.*, 1999). The package has been prepared primarily to help assess the quality of the EISs by review committees in the UK (1988-1990) and European Union countries (1996), and also in studies done in other countries (Lee *et al.*, 1999). A general checklist within the package allows some standardisation of the quality control of EIS, and may be applied in other countries, including Indonesia.

Use of the package

The package consists of three components:

- a guidance on how to use the package,
- a list of criteria to be used in each of the reviews, and
- a collation sheet to record the findings.

The guidance shows how to assess EIS, with criteria arranged in a hierarchical structure consisting of 4 steps (figure 4.5). First is the review, which draws upon assessments at the lowest level, the assessment of sub-category area (level 1). At the next higher level is assessment of category area (level 2). After applying more complex criteria to broader tasks and procedures in the process, the assessment moves to a higher level, the assessment of the review area (level 3), and the final step is application of the assessment to the overall quality assessment area (level 4). The results are recorded on a collation sheet using letters as symbols to discourage reviewers from using crude aggregation to obtain assessments at higher levels. The symbols represent values of assessments, as given below:

A. Generally performing well with no important tasks left incomplete.
B. Generally satisfactory, only minor omissions and inadequacies.

C. Satisfactory, despite omissions and/or inadequacies.
D. Unsatisfactory because of omissions or inadequacies.
E. Not satisfactory, with significant omissions or inadequacies.
F. Very unsatisfactory, with important tasks poorly done or not attempted.
NA Not applicable, as the package cannot be used.

Figure 4.5 shows assessment procedure for EIS review used in this research.

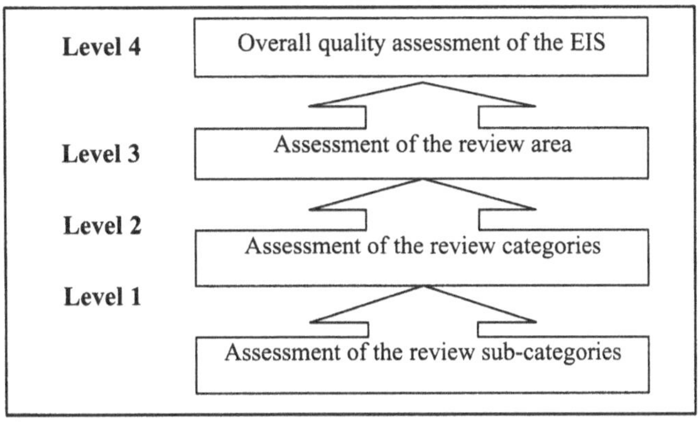

Figure 4.5. Assessment procedure for EIS review.

Selection of coastal EISs for the case study
The Sea Communications and the Industry offices received a total of 60 EISs for the District of Serang area during 1995-2000. Of these, 10 coastal EISs from each sector were selected first. Based on variation of use of jetty for the Sea Communications reports and variation of industry for the Industry reports, 5 reports per sector were selected for review in this research. The list of the coastal EIS reviewed in this research is given in table 4.2.

Table 4.2. The coastal EISs reviewed in the research.

Title of project of the coastal EISs	
The Sea Communications sector	Year
1. Special jetty development, Terminal for Liquid Material, Gurem, Pulo Merak Subdistrict	1999
2. Coastal Reclamation, PT. Hasta Mitra, Margagiri, Bojonegara Subdistrict	1997
3. Special jetty development, PT Cilegon Fabricators, Argawana, Bojonegara Subdistrict	1999
4. Special jetty development, PT. Indochlor Prakarsa, Mangunreja, Bojonegara Subdistrict	1999
5. Special jetty development, PT.Gt Petrochem Industries, Mangunreja, Bojonegara	1998
The Industry sector	
6. The Ethylene Glycol, PT. GT Petrochem, Mangunreja, Bojonegara Subdistrict	1998
7. Styrene Monomer Industry, Mangunreja, Bojonegara Subdistrict	1994
8. Coal Factory, Margagiri, Bojonegara Subdistrict	1994
9. Jababeka Industrial Estate, Kramatwatu Subdistrict	1998
10. Sugar Refinery Industry, PT. Bermis Madu Sejati, Bojonegara Subdistrict	1997

Note: Year indicates the year the reports were approved by the Minister for Communications or the Minister for Industry.

Procedure of the coastal EISs review in the case study

As suggested by the report on the EISs review (Lee *et. al.*, 1999), two experts should conduct the review in order to avoid a single overview and evaluation. So any significant differences in the assessment of individual review topics can be systematically examined and resolved before finalisation in the joint collation sheet. For this research, two experts were selected to review the coastal EIS, based on their knowledge of the EIA. They were familiar with EIA regulations, methodologies and their practice, and able to control the quality of the reviews. They were members of technical teams involved in EIS work on other projects.

Application of the method

For this research, the review topics were adjusted to the Indonesian EIA guidelines, and some sentences were simplified and adjusted to the Indonesian EIA terms in order to be more comprehensive. Training of the two experts and a trial of the evaluation with some revision on some sentences were conducted prior to the review.

The coastal EIS reviews in this research were conducted in 2 rounds. First, expert A reviewed the 5 reports from one sector while expert B reviewed the 5 reports from the other sector. The reports and review results were given to the researcher for an evaluation. After evaluating the reviews concluded by them and revising their methods as necessary, the two experts conducted the second round of reviews of the other 5 reports as well as revisions. In order to avoid bias in aggregating assessment values (denoted by symbols) of the reviews, as elaborated above, the experts were asked only to assess and give symbols to the sub-categories area, without aggregating the results. In order to make an average result of each category review area, the researcher translated the assessment symbols to numbers: 6 for symbol A, 5 for symbol B, 4 for symbol C, 3 for symbol D, 2 for symbol E, 1 for symbol F and 0 for symbol NA. Furthermore, the researcher aggregated the numbers in all sub-category areas in order to obtain an average score of the category review area. Finally, in order to have an average score as the overall assessment, the scores in all category review areas were aggregated.

Procedure of analysis in the case study

Analysis of results was conducted based on assessments of all reports in each sector, as well as of dominant cases. The dominant cases are the cases of the reports that have potential to generate environmental impacts and held public attention. Each report has 2 scores based on each expert's assessment, from which an average score is taken for each report. Scores for all the reports were grouped together in order to determine the average quality of the overall reports. The scores were then grouped according to sector in order to identify the quality of reports based on sector. For further assessment, analysis was focused on two selected dominant cases per sector. The dominant cases of the Sea Communications sector were report no. 1 (special jetty development) and report no. 2 (coastal reclamation). For the Industry sector, dominant cases were report no. 9 (Jababeka industrial estate) and report no. 10 (the Sugar Refinery Industry).

Sensitivity analysis
A sensitivity analysis of the score of each report was conducted by means of weighted summation and the Analytic Hierarchy Process (AHP) in order to identify ranks of quality for the assessment areas. A weighted summation based on scale of ranks given for each assessment area was applied to the scores of the EISs reviewed. An expert of the Banten Bay Research Programme who is responsible for the development of Banten Bay Information System (BBIS) and a representative of the district government who is involved in EIS review ranked each assessment area. The ranks of scale range from 7 (the highest) to 1 (the lowest rank). The ranks are averaged for further summation. The Analytic Hierarchy Process (AHP) for sensitivity analysis was also applied. The AHP method was used because it represents multi-criteria decision-making based on a hierarchy process using pairs-comparison. A representative of the district government who was dealing with environmental monitoring compared the importance between each assessment area, and gave ranks of scales of importance based on availability of environmental information and the use of data for coastal monitoring. A scale of 1/1 to 1/3 for the rank indicates "less important," and 1 to 3 for rank indicates "more important." The given ranks then were used as input for the AHP computer program in order to calculate averages of weighted scales for each assessment area. Table 4.3 presents ranks of scale of importance for assessment areas as discussed before. The scores were input to the AHP computer program, and the results are as shown in table 4.4. Summation of the AHP scores with the weighted scales to obtain total weighted ranks, as shown in table 4.5, was used for sensitivity analysis of the assessment areas.

Table 4.3. Ranks of scale of importance for assessment areas of the coastal EISs reviewed.

Areas of assessment	**Availability of information**				**The need for monitoring data**			
	A1	A2	A3	A4	A1	A2	A3	A4
A1. General description		3	3	1		1	1/3	3
A2. Evaluation of key impacts			1/3	1			1/3	1
A3. Mitigation and monitoring				3				3
A4. Communication of results								

Note: In the comparison between area 1 (general description) and area 2 (evaluation of key impacts) under availability of information, area 1 was considered as more important compared to area 2 (score was 3). Between area 2 and area 3 under availability of information, area 2 was less important compared to area 3 (score was 1/3).

Table 4.4. Scale of ranks for assessment areas of the coastal EISs reviewed.

Areas of assessment	**Scale of ranks**		
	Availability of information	**The need for monitoring data**	**Average**
A1. General description	0.407	0.223	0.315
A2. Evaluation of key impacts	0.119	0.162	0.141
A3. Mitigation and monitoring	0.295	0.487	0.391
A4. Communication of results	0.179	0.127	0.153

Table 4.5. Total weighted ranks of the coastal EISs reviewed.

Areas of assessment	Weighted scales			AHP	Total
	MIS expert	**Representative of district government**	**Average of ranks**	**Scale of ranks**	**weighted rank**
A.1. General description	7	7	7	0.315	7.315
a. Description of the plan	7	7	7	-	7
b. The affected environment	4	1	2.5	-	2.5
c. Baseline conditions	1	2	2.5	-	2.5
A.2. Evaluation of key impacts	1	5	3	0.141	3.141
a. Scoping	5	3	4	-	4
b. Describing of key impacts	7	5	6	-	6
c. Assessment of impacts	1	7	4	-	4
d. Sustainability of the plan	3	1	2	-	2
A.3. Mitigation and monitoring	5	3	4	0.391	4.391
a. Alternative measures	7	3	7	-	7
b. Mitigation measures	5	7	5	-	5
c. Monitoring measures	3	3	3	-	3
d. Recommendations	1	1	1	-	1
A.4. Communication of results	3	1	2	0.153	2.153
a. Layouts	3	7	5	-	5
b. Presentation	7	5	6	-	6
c. Uncertainties	5	4	4.5	-	4.5
d. Emphasis	4	3	3.5	-	3.5
e. Consultation	2	2	2	-	2
f. Non-technical summary	1	1	1	-	1

4.4.4. Assessment of environmental information sources to draft and review EIS

Research question 3: What environmental information sources are used by consultants to draft EIS and by the technical teams and the district governments to review it?

The quality of the EIS also depends on the use of information sources to draft and to review it. "Use" of the information sources here, in a broader sense, refers to how information is "obtained and used," which also covers the "need and availability" of information sources (Brown 1994). A better understanding of the need for, availability and use of information sources enhances the role of district governments in the EIS review. Insight into information sources gives the technical teams, the key players involved in the EIS review, a better perception of the environmental information that can be used to support analysis of the research in the EIS.

The aim of this research is to review the use of environmental information sources by consultants to draft the EIS and by technical team and district government to review the EIS. "Use" of the information sources here, in a broader sense, refers to how information is "obtained and used," which also covers the "need and availability" of information sources.

Development of questionnaire

A questionnaire that covers 17 types of sources of environmental information was developed in order to get an overview of the quality of information sources related to the EIS. Fifteen of these sources were adopted from a study report titled, "Acting locally: Meeting the environmental information needs of local government" (Brown 1994). These 15 information sources have been gathered from various surveys in Australia according to the needs of information sources for local governments there. The questions were altered to conform to Indonesian terms of information usage for the EIS. Two types of information sources, status of natural resources and environmental monitoring, were added to the questionnaire, as these sources are important for coastal management. The questionnaire consists of two main groups of questions:

- Purpose of the EIS.
- Use, need, the availability, and the accessibility of information sources. This group consists of 2 subgroups of information sources: sources related to the EIA data, and flow of information or networking.

Try out of questionnaire

One representative of each group of respondents was given a "trial" questionnaire and accordingly sentences were changed without disturbing the content in order to improve the clarity.

Selection of respondents

Respondents for this research consisted of three groups related to the EIS:

- Members of EIA technical teams who review the EIS from the Industry and the Sea Communications offices and from other sector offices.
- Representatives of consultants who, together with developers, prepare the EIS for the Sea Communications and Industry offices.
- District governments of the study area that should be involved in the EIS review of the study area.

Distribution of questionnaire

The questionnaire was distributed to the focus groups of respondents related to the coastal EIS, the consultants, the technical teams, and district governments. The distribution was done in three ways:

- Questionnaires were distributed during coastal EIS review meetings of the Industry and Sea Communications sectors.
- Questionnaires were faxed to members of the technical team of the two sectors and of related sectors like Forestry, Internal Affairs and Environment, and to the consultants' offices.
- Questionnaires were distributed directly to district government offices such as District Planning, District Environment, District Mining, District Fishery, District Industry, and District Port Authority.

The researcher was present at district government offices to help in completing the questionnaire by explaining to respondents of the district government the terms of environmental information sources that they were unfamiliar with.

Answers to the questionnaire were received by fax or handed over to the researcher directly (for example, during the observation of the coastal EIS review meetings). The total number of questionnaire forms distributed was 71 (100%), and the number completed and returned was 59 (82%), which can be considered as a high percentage return. Table 4.6 presents distribution of questionnaire and returns in this research.

Table 4.6. Distribution of questionnaire and returns in the research.

Consultants		**District governments**		**Technical Teams**		**Total**	
Out	Return	Out	Return	Out	Return	Out	Return
20	17	16	14	35	28	71 (100%)	59 (82%)

A total of 59 respondents participated in this research, of the 59:

- 12 represented the 27 members of the technical team of the Sea Communications and the Industry sectors, and 16 members of the technical teams outside these two sectors,
- 17 represented the 35 EIA consultant offices registered at the Sea Communications and Industry offices, and
- 14 were from the district government offices related to activities of the Sea Communications and Industry sectors.

4.4.5. Observation of the coastal EIS review meetings

Research question 4: How is environmental information used in the coastal EIS review meetings?

The EIS review is undertaken primarily to determine the quality of environmental information presented in the EIS as a basis for its approval. The coastal EIS review meetings in this study are based on decree no. PP 51/1993 in which conducted at national level, because the Ministerial Decree for Environment no. 41/2000 on the establishment of review committee for district level was not yet implemented when this research was carried out. According to decree no. PP 51/1993, the EIS has to be submitted to the central EIA commissions of the responsible sector, that is the review committee. The review is then formally carried out at national level with the help of the technical team. The aim of this research is to assess how the environmental information in the coastal EIS is evaluated in the coastal EIS review.

Selection of coastal EIS review meetings

Of the meetings conducted by Sea Communications and Industry sector offices during 2000, most dealt with EIA cases that are outside the focus of this research. They were not related to either the study area or coastal management activities. The Sea Communications cases were mostly conducted in outside islands that are geographically different to the Java Island where the study area is located. Two cases from each sector of activities were selected after discussion between the researcher and the central commissions of the two sector offices. The four cases selected for the review process were considered as sufficient for the purpose of getting an overview of the review process. Observations of their EIS reviews were done with the permission of the secretary of the technical teams. The dates and particulars of the selected coastal EIS review meetings observed in the research are given in table 4.7.

Table 4.7. Coastal EIS review meetings observed in the research.

Case	Sector and case of the coastal EIS review meetings	Project status	Date of observation
	The Industry office:		
Case 1	Industrial Estate development in Tangerang, Banten province	Part of area is in operation	9 February 2000
Case 2	Paper Industry development, Lampung province	In operation since 1977	19 March 2000
	The Sea Communications office:		
Case 3	Dredging for Sea Shipping in Moro Sea waters, Karimun, Riau Province	In a development plan	20 July 2000
Case 4	Dredging for Sea Shipping in Durian & Sugi Straits, Karimun, Riau Province	In a development plan	21 July 2000

Note: Project status is the status of the project when the EIS was reviewed.

Case 1 of the Industry office was selected because it is similar to a case that exists in the study area. Case 2 of the Industry office was selected due to its proximity to the study area. The Sea Communications cases were selected because the issues under consideration were similar to the problems created by sea sand mining activities in the study area. The Director General of Sea Communications of the Ministry of Communications issued the permits for these two projects although the Ministry of Mining usually does it. This was because the two projects were for sea shipping purposes and is related to the safety of sea transportation lines, which is an uncommon situation in Indonesian EIA cases.

During 2000 (January-November 2000), on average, only 2 coastal EIS review meetings per month were conducted in the EIA central commission of the Sea Communications office. And only 8 EIS review meetings were conducted in the Industry office at national level for the same year. The observations on four meetings for both the offices are considered as highly representative for the coastal EIS review meetings.

Development of review criteria

A list of review criteria was developed in this research as a guideline for the observation of coastal EIS review meetings, mainly to assess discussions on environmental information in the coastal EIS review meetings. The review criteria consist of groups of sub-criteria:

- General overview of time used in the meeting for: preparation of attendants to the review process, presentation of the coastal EIS document, and discussion of the contents.

- Discussion on the contents of the coastal EIS: description of the plan, baseline conditions, and assessment of key impacts, mitigation measures, and monitoring measures.
- Criteria used for time spent for presentation of the document: > ½ hour, < ½ hour, and none, and time spent for discussion: > ½ hour, < ½ hour, and none.

Implementation of the review process

The researcher was present in all review meetings and filling out the review checklist as well as taking note on the observations of the review processes.

4.4.6. Assessment of environmental monitoring

Research question 5: What environmental parameters and indicators are monitored by the district government?

One of the uses of the EIA in the decision-making process of EIS approval is to implement impact-monitoring measures (UNEP, 1995). Therefore, the use of the environmental information in the EIA process can be determined from the quality of the environmental monitoring activity. Although the EIS is approved at national level, the implementation of decisions on impact mitigation and impact monitoring are at district level. The separation of institutions that approve and monitor the implementation of EIS decisions allows interventions on the quality of the environmental monitoring. The aim of this research is to analyse the monitoring of environmental parameters and indicators by the district government in general.

Development of questionnaire

The questionnaire covers parameters and indicators to monitor the environment and environment impacts. A list of parameters was compiled from the technical guidelines for the Sea Communications and Industry sectors, as well as from the Indonesian general guideline. Parameters for environmental monitoring were grouped into 5 environmental components:

- hydrology and river water quality,
- hydro-oceanography,
- spatial, land and soil,
- flora and fauna, and
- socio-economic and culture

Parameters for environmental impact monitoring were grouped into 7 components related to changes of the above environmental components. There were 3 choices answers on frequency to questions on the use of parameters for environmental monitoring and impact monitoring: *often, less often, and never. Often* means the use of indicator for environmental monitoring is in accordance with the ideal frequency. Alternative answers to constraints in the use of parameters were given under 4 choices: *laboratory facilities, human resources, data recording and analysis, and financing.*

Selection of respondents

Four respondents, two staffs out of 20 members of the Planning office and two staffs out of 15 members of the Environment office in the District of Serang in the Banten Bay area were selected to be the respondents, because their offices are important in the implementation of environmental monitoring. The selection of 4 four respondents is considered as highly representative for the assessment of environmental monitoring conducted by the district government of Serang.

Implementation of the methodology

Questionnaires were given directly to the respondents along with explanations of the questions. The researcher was present to help in filling checklists and prevent bias of misinterpretation of questions. Some comments from the respondents revealed that they were not familiar with all the parameters in the checklist because they use only a selection in their activities. They found it difficult to differentiate between parameters for environmental monitoring and those for environmental impact monitoring. Some of the answers were related only to their daily activities. "Never" meant that another district institution did the monitoring, or that it was not being done in the field, or that they were not sure what the question meant. Sometimes the answer given as "less often" really meant "never."

4.4.7. Assessment of perceptions of technical team on the EIA process

Research question 6: How is environmental information used in the EIA process based on perceptions of the technical team?

The previous research questions will provide an overview of how environmental information is being used in the assessment and practice of the Indonesian EIA process in the coastal zone based on decree no. PP 51/1993, in which the EIS review is conducted at national level. These results, combined with the assessments of perceptions of technical teams at national level, will be used to suggest instruments that can be used to enhance the Indonesian EIA process, in the coastal zone in general and in the case study in particular.

The aim of the assessment is to generate information on the perceptions of the technical team on:

- contents and use of EIA technical guideline,
- quality of the EIS,
- use of information sources, and
- the EIA process.

An open-ended questionnaire on the perceptions of the technical team was used for this research so that information generated can easily be structured while respondents still have opportunities to express opinions on matters that are outside the checklist.

Development of questionnaire

A list of questions concerning the research objective was developed in order to review perceptions of the technical teams on the technical guidelines. Questions concerning stages in the EIA processes and key parties responsible were identified based on the EIA general guideline. Questions on aspects of use and benefits of technical guidelines were developed based on a reference review (Sadler and Verheem, 1996). Questions on environmental components and parameters were taken from the Indonesian EIA technical and general guidelines, with some adjustments.

Selection of respondents

Two respondents at national level, a secretary and a member of the technical teams were selected for in-depth interview from each of the sectoral offices of the Sea Communications and the Industry. They are important stakeholders for quality improvement of the EIA process. In order to have an independent opinion or perception, an advisor of the State Minister for Environment was also asked to fill in the checklist. The selection of two respondents out of 15 members of technical team of the Sea Communication sector, and two respondents out of 12 members of technical team of the Industry sector is considered as sufficient.

Distribution of the questionnaire

Questionnaires were sent to the 5 respondents by fax, and collected a few days later. Questions were clarified through communication by telephone between the researcher and the respondents.

Implementation of the research

Only 4 of the 5 questionnaires were filled and returned. The respondent who did not return the questionnaire still holds the position, secretary of a technical team. He has moved to another section of his office and was difficult to contact. The respondent from the State Ministry for Environment commented on stages in the EIA process. Other comments received are related more to the expectation of a more open-ended questionnaire, and opinions on the practice of the EIA.

4.5. Research Design: Stakeholder Participation in the coastal EIA Process

4.5.1. Objectives and research questions

Objectives

As discussed in chapters 2 and 3, the quality of stakeholder participation in the coastal EIA process is very important. Therefore, this research aims:

1. to assess the participation of stakeholder in the coastal EIA process at the district level, and
2. to provide suggestions for improving the stakeholder participation based on research results and potencies and prospects of new regulations.

Stakeholder participation in the coastal EIA process in this research is assessed on the basis of the perceptions of the technical team and the quality of participation in the coastal EIS review meetings. An illustration on the practice of stakeholder participation in the pre-coastal EIS compilation is presented based on a case study. The research was focused on laws on stakeholder participation that preceded the new regulations, which are used in analysis of the research results.

Figure 4.6 shows studied elements within stakeholder participation used in this research that will be elaborated in chapters 6. Elements of stakeholder participation in the coastal EIA are legislation on stakeholder participation and stakeholder participation needs.

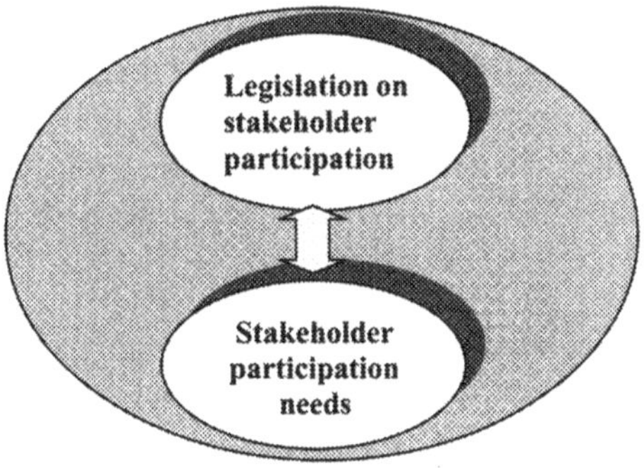

Figure 4.6. Studied elements of stakeholder participation in the research.

Research questions

To realise the objectives of this research, the following questions were developed.

1. How is the participation of stakeholders in the EIA process based on perception of technical teams?
2. How is the participation of stakeholders in the coastal EIS review meetings?
3. How is the participation of stakeholders in the pre-coastal EIS compilation as illustrated in the case study?

A summary of the methodologies used in the assessment of stakeholder participation is shown in table 4.8.

Table 4.8. Summary of the methodologies used in the assessment of stakeholder participation in the coastal EIA process.

Research questions	Methods	Samples/Respondents
1. Assessments of perception of technical teams on stakeholder participation	Questionnaire	4 respondents at the national level
2. Observation of stakeholder participation in the coastal EIS review meetings	Checklist	4 meetings at the national level
3. Assessments of stakeholder participation in the pre-coastal EIS compilation as illustrated in the case study	Questionnaire	89 respondents at the district level

Methodological Procedure

The participation of stakeholder in the coastal EIA process in this research is assessed on the basis of perceptions of technical teams on stakeholder participation, the participation of stakeholder in coastal EIS review meetings, and in the pre-coastal EIS compilation process. Stakeholder participation in the pre-coastal EIS compilation, for example in project approval, is used to illustrate the practice of stakeholder participation in the coastal EIA process.

The participation of stakeholder in the EIA process is assessed based on the perception of the technical team on participation of stakeholder, regulations and practice, and perception of their familiarity with the new concept of stakeholder participation. The participation of stakeholder in the coastal EIS review meetings is assessed based on observation of stakeholder participation in the meetings, and inputs generated by the participants to the meetings. The participation of stakeholder in the pre-coastal EIS compilation is assessed based on perceptions of stakeholders on their participation in coastal development approval, their communications with the project, and their contribution of inputs or information to the project plan. Based on the conclusion of the results assessment and prospects on implementation of decree no. 08/2000 on stakeholder participation, suggestions for improving stakeholder participation in the Indonesian EIA process in the coastal zone are proposed.

Focus of the research

The research focused on two sectors, the Sea Communications and the Industry offices at national level as the main sectors in the study area (see chapter 3), and the local stakeholders in the District of Serang as the case study area. All research for this chapter focused on the practice of stakeholder participation before the passing of Ministerial Decree for Environment no. 41/2000 on the formation of the EIS review committee at city/district level and decree of chairman of Bapedal no. 08/2000 on public participation in the EIA process. The two regulations are used for comparison in the discussion of the research results. In this research, the terminology used is stakeholder participation, which refers to the public in general and the district government, as well as to the process where stakeholders can influence the decision-making of the coastal EIA process.

4.5.2. Assessment of stakeholder participation in the EIA process based on perceptions of technical teams

Research question 1: How is the participation of stakeholders in the EIA process based on perception of technical teams?

The aim of the question is to generate information on the perceptions of the technical team on the regulation and the practice of the EIA process in general on stakeholder participation. The technical team plays an integral part in the EIA process, as it is almost continuously involved in all steps of the EIA process, from screening to approval. Therefore, the perceptions of the technical team on the participation of stakeholder as in the law and in practice is important for the evaluation of the EIA process. This assessment also provides insight into the familiarity of the technical team with the new concept of stakeholder participation in the EIA process as defined by decree no. 08/2000. This research is based on decree no. PP 51/1993, in which the review committee established at national level.

An open-ended questionnaire was developed. It contains a list of main questions on perceptions of respondents on stakeholder participation as in regulations and used in the practice related to:

- stages of the EIA process and groups of stakeholders involved,
- responsible groups of stakeholders involved for each stage, and
- participation of stakeholders in the EIA process.

Information concerning methods used for this research on the development of the questionnaire and selection of respondents are the same as that in section 4.4.

4.5.3. Observation of stakeholder participation in the coastal EIS review meetings

Research question 2: How is the participation of stakeholder in the coastal EIS review meetings?

Decree no. 08/2000 on stakeholder participation in the EIA process specifically defines the procedure for stakeholder participation in EIS review meetings. Decree no. 41/2000 on the establishment of review committee at district/city level stipulates the approval process of the EIS within the local government level. Since the two decrees were not implemented when the research was carried out (1999-2000), there is concern regarding the participation of stakeholder based on the previous law that defined the EIS review process and the EIS approval within the national level of government sector. The aim of the question is to assess the conduct of coastal EIS review meetings, particularly, in relation to the role played by stakeholders in the coastal EIS review process.

A review checklist was developed as a guideline for the observation of coastal EIS review meetings. The checklist consists of review items on participation of stakeholders during the meetings and in the review process. The main review items of the checklist were:

- attendance at the meetings,
- participation of attendants in giving questions and comments, and
- approval process of the documents, including:
  - input from attendants to the meeting
  - perceptions of agreements reached by the attendants,
  - final agreement reached in the meeting.

Information concerning methods used for this research on the development of questionnaire, try out and distribution of questionnaire, and selection of respondents are the same as those in section 4.4.

4.5.4. Assessment of stakeholder participation in the pre-coastal EIS compilation

Research question 3: How is the participation of stakeholder in the pre-coastal EIS compilation as illustrated in the case study?

Stakeholder participation in the pre-EIS compilation, such as the planning process, is increasingly recognised as an important component addressed by the new decree no. 08/2000 on stakeholder participation. Therefore, it is important to assess the participation of the stakeholder before the new decree was passed. The aim of this research is to assess the conduct of stakeholder participation in development approval, as illustrated in a case study. As illustrated in the case study of the Jababeka industrial estate, the practice of stakeholder participation at district level in the pre-coastal EIS compilation stage is reviewed.

Selection of a case study

The case study used for this research is the Jababeka Industrial Estate project, located in the Banten Bay of the District of Serang. The project plan covered an area of 2,185 hectares, most of it has been owned by the project. A part of this land area is used mainly for agriculture, including 600 hectares of fishponds. Another part will come from sea reclamation activity (615 ha). The land area of the project is within 2 village administration areas, Terate village and Tonjong village. Both these villages come under the administration of the Kramatwatu sub-district. Most fishermen use about 800 hectares of seagrass, located close to the project area, as a fishing ground. Permits for reclamation and jetty development were granted by the Minister for Communications through the Director General for Sea Communications while the Minister for Industry issued permits for development of the industrial estate. The former are not approved yet (1999) while the latter were approved in 1999. Due to a monetary crisis, the project is under the authority of the central government as the investor has gone bankrupt.

Development of questionnaire

Part of the questionnaire dealt with issues that might arise from the development of such an environmentally sensitive industrial estate covering such a large area (2,185 hectares), 615 hectares of which was claimed from the sea. Another part of the questionnaire was based on a literature review (Doyle, 1999). The questionnaire aims to find out:

- the perceptions of stakeholders on their participation in project approval,
- their access to information on the project, and
- their perceptions of the performance of the project.

Selection of respondents

Respondents to the survey were the coastal community within the study area and officials related to the coastal project at various levels of district government in the District of Serang. Respondents were selected by non-random sampling in order to obtain information on the general perceptions of stakeholders. The focus groups of coastal respondents included farmers, pond owners and pond labourers, fishermen, and informal sector/small business owners. The district government officials were sub-district and village authorities of the sectors of district government that are involved in development approval (such as District Mining and District Fishery offices), and representative of the Planning and Environmental offices. The total numbers of respondents (86 people) participating in this survey was considered as sufficient, as they were representing small populations within the coastal study area (1,570 hectares), where most of the land use previously was for agriculture and fish farming.

Try out of questionnaire

A trial questionnaire was conducted with representation from each of the focus groups in order to find out their ability to comprehend the questions. Accordingly some questions were simplified without changing the content.

Distribution of questionnaire

The questionnaire was distributed to the focus groups of respondents with the assistance of 10 students of the Institute of Fisheries (*Akademi Perikanan Laut*) in the area, under the supervision of one of their teachers. They were accompanied by a local guide who was familiar with the respondents of each coastal village they visited and who helped in the translation of the questionnaire into the local language. The students visited a coastal village together, meeting farmers, pond owners, and pond labourers in their fields during the day and fishermen in the beach during the evening. The students were trained on the conduct of the survey in order to prevent misinterpretation of questions. The researcher conducted the questionnaire survey among district government officials in order to get more insight into the information.

A total of 100 questionnaires were distributed to the respondents, in which 25 questionnaires distributed to the district government officials and 75 questionnaires to the selected coastal community respondents. A total of 86 respondents participated. They consisted of:

- 18 representatives of all levels of district government in the District of Serang.
- 68 people of the coastal community: 14 farmers, 10 pond owners, 15 pond labourers, 20 fishermen, 5 middlemen or broker of the fishermen, and 9 representatives of small business such as local restaurants, rice mills, and dry-fish trade.

The distribution of respondents who participated in this research is shown in table 4.9.

Table 4.9. Distribution of focus groups of respondents in the research.

Group of respondents	**District government**			**Coastal community**					
Area	District	Sub-district	Village	Farmers	Pond owners	Pond labourers	Fisher - men	Broker of the fishermen	Small business reps.
District of Serang	8								
Sub-district of Kramatwatu		2							
Village of Terate			4	8		10	20	5	5
Village of Tonjong			4	6	5	5			4
Total (86)	8	2	8	14	5	15	20	5	9

PART III

ANALYSIS OF THE CASE STUDIES

5.
USE OF ENVIRONMENTAL INFORMATION IN THE COASTAL EIA PROCESS

5.1. Introduction

This chapter presents an analysis of the use of environmental information in the coastal EIA process. The analysis is conducted based on the following assessments:

- environmental information stated in the EIA guidelines (section 5.2),
- environmental information described in the coastal EISs (section 5.3),
- information sources used by consultants to draft EIS (section 5.4),
- information sources used by technical teams and district governments to review EIS (section 5.5),
- environmental information discussed in the coastal EIS review meetings (section 5.6),
- environmental information in the environmental monitoring (section 5.7), and
- perceptions of technical teams on use of environmental information in the EIA process (section 5.8).

5.2. Environmental Information in the EIA Guideline

Three EIA guidelines came under in-depth review in this research. They are:

1. the technical guideline for the Industry sector,
2. the technical guideline for the Sea Communications sector, and
3. the general Indonesian EIA guideline as the source of the two technical guidelines.

Each guideline consists of checklist of components of the environment and the environmental impacts to be taken into account in the EIA study. Each component consists of parameters, each of which consists of several indicators. The sufficiency of indicators in each parameter was considered in the in-depth review of this research (section 4.4.2). A parameter is considered by the experts as sufficient if it has a sufficient number of indicators that can describe the quality of the said parameter.

5.2.1. EIA technical guidelines of the Industry sector

The environmental components listed in the technical guidelines of the Industry sector consist of spatial, physico-chemical, biological and socio-economic components, with a total of 15 parameters. Based on the review by experts of indicators for each parameter, 7 out of 15 parameters are reported to have sufficient indicators, which consist of all three parameters (demography, socio-economic, and socio-culture) of the socio-economic component and four parameters (climate, physiography, hydrology, and hydro-oceanography) of the physico-chemical component. Almost all parameters within the spatial and the biological components have been reviewed as having relatively insufficient indicators to describe the baseline condition. The environmental impact components in the technical guideline of the industry sector are the same as those of the environmental component given above. A total of 13 parameters have been reported, of which 10 out of 13 parameters are reported have sufficient indicators while 3 parameters had insufficient indicators. The latter consist of one parameter (hydrology) of the physico-chemical component and all parameters (land and waters biotic) of the biological component.

There is no technical guideline available for the impact mitigation and monitoring measures in the Industry sector. The available guideline is in the form of a blank checklist, which needs to be filled in with relevant indicators for impact mitigation and impact monitoring measures, at the same time as when the EIS is being proposed. The guideline includes only executing institutions for impact mitigation while no institutions are mentioned for impact monitoring. Table 5.1 shows a summary of the results of the EIA technical guideline of the Industry sector reviewed by experts. Addition of indicators suggested to the EIA technical guideline of the Industry sector is presented in Appendix I.

5.2.2. EIA technical guideline of the Sea Communications sector

The environmental components within this technical guideline consist of land, water resources, hydro-oceanography, air and sound, flora and fauna, socio-economic and "other" components, with a total of 23 parameters. Based on the review by experts of the indicators within each parameter, 17 out of 23 parameters are reported as having insufficient indicators to describe the baseline condition. The only component where parameters have sufficient indicators is the hydro-oceanography component that relates to the main activity of the Sea Communications sector. The environmental impact component in this technical guideline, with a total of 17 parameters, consists of physico-chemical, biological, and socio-economic components, all of which have insufficient indicators.

All parameters for the impact mitigation and monitoring measures in this technical guideline are reported to have sufficient indicators. This guideline does not mention institutions for the implementation of impact mitigation and impact monitoring measures. Table 5.1 shows a summary of the results of the EIA technical guideline of the Sea Communications sector reviewed by experts. Addition of indicators suggested to the EIA technical guideline of the Sea Communications sector is presented in Appendix II.

5.2.3. Indonesian EIA general guideline

This general guideline consists of four environmental components, physico-chemical, biological, socio-economic and public health components, with a total of 16 parameters. The experts report that all parameters within these components have sufficient indicators, except the biological component, where most parameters have insufficient indicators to describe the baseline condition. There are four components within the environmental impact components, with a total of 16 parameters. Here too all parameters within the components other than the biological component are reported to have sufficient indicators.

Concerning impact mitigation and impact monitoring, out of 12 parameters identified for impact mitigation and impact monitoring measures in this guideline, 9 are reported to have sufficient indicators. This general guideline only mentions executing institutions for impact mitigation, while executing institutions for impact monitoring are not stated.

Table 5.1 shows a summary of the results of the Indonesian EIA general guideline reviewed by experts.

Table 5.1. Summary of the results of the EIA guidelines reviewed by experts.

Components of the guidelines	**Technical guidelines** of the Industry sector	**Technical guidelines** of the Sea Communications sector	**The Indonesian EIA general guideline**
Environmental components	7 out of 15 parameters have sufficient indicators. All parameters in the spatial and biological components have insufficient indicators	17 out of 23 parameters have insufficient indicators. The hydro-oceanography component has sufficient indicators	All components except biological component have sufficient indicators
Environmental impact components	3 out of 13 parameters have sufficient indicators. All parameters within biological component have insufficient indicators	All parameters have insufficient indicators	All parameters in the biological component have insufficient indicators
Impact mitigation and monitoring measures	The available guideline is in the form of a blank checklist.	All parameters have insufficient indicators	All parameters in the biological component have insufficient indicators
Institutions for impact mitigation: *executing institutions* *monitoring institutions* *related institutions*	Only executing institutions stated in the guideline	No executing institution stated in the guideline	Only executing institutions stated in the guideline
Institutions for impact monitoring: *executing institutions* *monitoring institutions* *related institutions*	Only executing institutions stated in the guideline	No executing institution stated in the guideline	No executing institution stated in the guideline

5.3. Environmental Information Described in the Coastal EISs

Based on the review assessment method, using review criteria as explained in section 4.4.3, the quality of the coastal EISs is presented in scores based on all reports, sectors, and the dominant cases. Each score consists of an average score for overall quality assessment and assessment areas. For further assessment, as for sensitivity analysis, total weighted ranks are applied as a summation of weighted-scale and analytical hierarchical procedure (AHP) analysis.

5.3.1. Quality of the coastal EISs reviewed based on all reports and sectoral reports

Score C is considered as the level of acceptable for the reports. From the 10 reports reviewed, 50% of the total reports reviewed in average have sufficient quality for the acceptance (C), while the rest of the reports have insufficient quality for the approval (E). The four dominant cases, which might pollute the coastal environment, obtained almost similar scores as all other reports, with an average score of D in all assessment areas and in overall assessment. As seen in figure 5.1 and Appendix III, the average score of reports is inadequate or below C (average score of all reports is D) in the overall assessment, as well as in all assessment areas. However, all reports have been approved in practice, although they require much improvement to reach the level of acceptance. The Sea Communications reports in average have score C and above, indicated the reports have sufficient quality for the acceptance. In contrast, the Industry reports in average have unsatisfactory quality for the approval, indicated by score E for the overall quality. Average scores of the Sea Communications reports on all review areas are higher compared to the average scores of the Industry reports. The average score for the Industry reports in the review area on general description (1) is higher (D) compared to other areas (E). Similarly, the average scores of the Sea Communications reports in the review area on general description and impact mitigation and monitoring measures are higher (B) compared to other review areas (C). In general, common features emerges, such as:

- Average reports within each sector tend to have sufficient quality only in the assessment area of general description but are weak in other areas such as evaluation of key impacts. This was probably because unlike in the assessment area where a more descriptive response was required while in the latter area more quantitative methods on environmental impact assessments were required.
- As seen in Appendix III, all reports tend to conform to better standards in description of the plan (C), compare to the effected local environment (D) and even the base-line conditions (E). This suggests that the consultants are more familiar with descriptive type of EIS compilation. The reports tend to be less satisfactory where quantified measures are required such as in description of the affected local environment, and even worse in description of the base-line condition where quantified measures and more environmental data is needed.
- In the analysis of the assessment area on evaluation of impacts, (Appendix III), all reports tend to have better quality in areas on environmental appraisals and description of key impacts while they are of poor quality in areas on assessment of

impacts (E). This would suggest that the former areas are more descriptive than the latter area that contains more rigorous tasks on environmental impact assessments.

- In the analysis of the area on mitigation and monitoring, (Appendix III), all reports are of poor quality in the sub area on monitoring measures (E) compared to other areas on alternative measures, mitigation measures, and recommendation (D). This figure would suggest that the weakest area in average of all reports as well as the Industry reports is the assessment of monitoring measures. In contrast, the Sea Communication reports are on average of has better quality in the sub area on monitoring measures (A) compared to other sub areas (B and C). Also, the Industry reports in the sub area on monitoring measures are of better quality (D) compare to the assessment of other sub areas (E).
- Based on the sensitivity analysis as presented in Appendix III, all reports and the 4 dominant cases have different average quality but have the same order of ranks of quality for each assessment area. The highest quality is found in the area of general description (1), while the lowest quality is found in the area of evaluation of key impacts (2). This finding supports the previous statement that the more descriptive assessment tends to be of better quality compared to the assessments that require more quantified measures and more tasks on environmental impact assessments.

Other researchers recorded similar trends, for example Lee *et. al.* (1999), and Lee and George (2000) in their assessment on the use of the review package for a number of samples of the European countries report that the assessment areas that contain more descriptive requirement tend to be of higher quality than the assessment area that contains more quantified measures. Similar trends are also indicated by Mwalyosi *et. al.* (1998) on the use of the review package for the assessment of a small number of Tanzanian EIA reports. Figure 5.1 shows the average quality of the 10 EISs reviewed differentiated for two sectors, the Sea Communication and the Industry sectors.

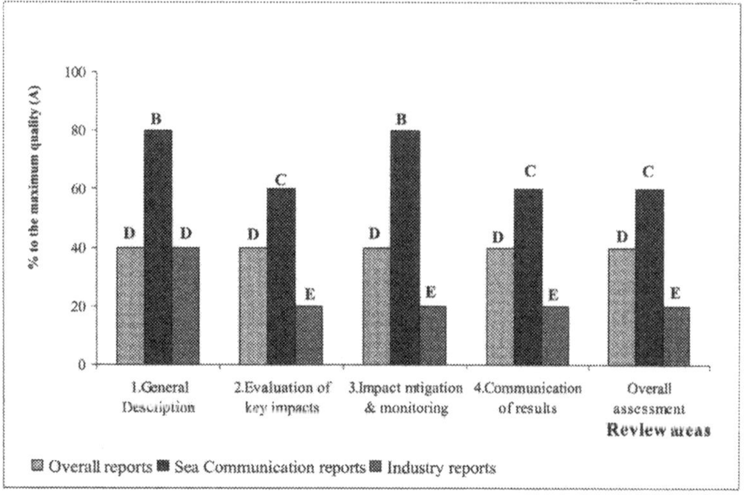

Figure 5.1. Average quality of the 10 EISs reviewed differentiated for two sectors.

5.3.2. Quality of the 10 coastal EISs reviewed based on sectors and two dominant cases per sector

Figure 5.2 shows the reports reviewed differentiated for each sector, the Sea Communication and the Industry, as well as two dominant cases per sector. Figure 5.2 and appendix III show that in the two dominant cases, the quality of the Sea Communications reports is, on average, higher (B) compared to that of the Industry sector (D). In the assessment of monitoring measures of sub-area (3.3), the two dominant cases of the Sea Communications reports have the highest score (A), which is considered useful for monitoring purposes, while the Industry reports are reviewed as having a low score (D). Average scores of the Sea Communications reports on review areas such as general description (1) and mitigation and monitoring (3) are higher (B) compared to other review areas. The average score for the Industry reports in the review area on general description (1) is higher (D) compared to other areas (E). The same pattern is seen in the dominant cases of both sectors. This means that for the dominant cases, the Industry reports need much improvement in its quality of acceptance compared to that of the Sea Communications reports. As seen in figure 5.1, all reports of the Sea Communications sector having a score of C and above are of higher quality than reports of the Industry sector, which have score D or below. The Industry reports need much more improvement in its quality to reach the approval score C. Similar to the assessment of all reports as in previous section; this assessment shows common features, such as:

- Reports of the dominant cases within each sector tend to be of better quality compared to the average quality of the reports of the Sea Communications and the Industry sectors. This suggests that in the compilation of the EIS for the dominant cases are better accomplished compared to the compilation of other reports. The reason for this maybe that more thorough research and thorough impact assessment would be conducted for the dominant cases and more guidance as well as assistance given by the technical team in directed the EISs compilation.
- Unlike the sectoral reports that have better quality only in the assessment area of general description but are weak in other areas such as evaluation of key impacts, reports of the dominant cases for each sector tend to contain the same quality in all of the assessment areas. Possibly, because the dominant cases commonly are big projects have sufficient funding to conduct the EIS compilation, so that all areas assessment was thoroughly assessed in the compilation of the EIS.
- Each assessment area of the dominant cases tends to be of have similar quality except the general description area of the dominant cases within the Industry sector, where the quality of the baseline condition sub area is the lowest (E) compared to other sub-areas. The situation is similar in all reports within the Industry sector. This may possibly be due to, localised environment data not being available to support the EIS compilation.
- Based on the sensitivity analysis as seen in Appendix III, both sectors have different average quality but have the same order of ranks of quality for each assessment area both for all reports as well as for the dominant cases. The highest quality is found in the area of general description (1), while the lowest quality is

found in the area of communication of results (4). This suggests that the areas that have qualified assessment tend to have higher quality compared with the areas that have quantified assessment.

- The area for impact mitigation and monitoring (3), which is very important for coastal monitoring, has the second highest quality for both sectors. This consists of 4 subcategory areas: alternative measures (3.1), mitigation measures (3.2), monitoring measures (3.3), and recommendations (3.4), in which monitoring measures has the second lowest quality for both sectors. This suggests that the sub area on monitoring measures has been considered as unimportant by the respondents.
- The reports of the Sea Communications sector are of higher qualities compared to those of the Industry sector before as well as after the weighted scales were applied. This suggests that the use of the review criteria is in accordance with the ranks of importance of the assessment areas given by the experts and the district government staffs for this research.

Figure 5.2 shows the average quality of the 10 coastal EISs reviewed differentiated for each sector, the Sea Communication and the Industry, as well as for two dominant cases per sector.

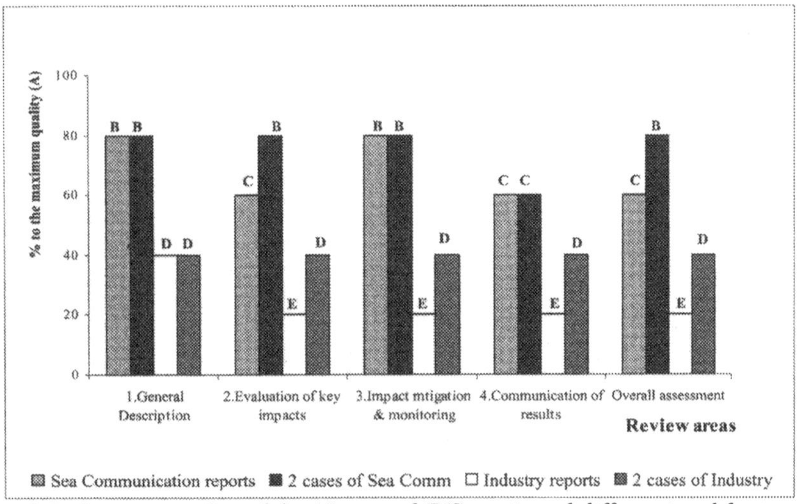

Figure 5.2. Average quality of the 10 coastal EISs reviewed differentiated for two sectors and two dominant cases.

5.4. Information Sources Used by Consultants to Draft EIS

A list of 13 types of environmental information sources that are most needed to draft EIS and 4 types of flows of information were compiled for this research and submitted to the consultant respondents. The sources of information used by consultants to draft EIS are assessed on usage, how they are being perceived as necessary, and their availability. Each of these is assessed in three scales, "most", "in between" and "not". As an example, the availability is assessed as mostly available, available and not available.

Discussions will first focus on the perception of consultants on the five most used, five most relevant, five most needed, and five most available information sources. Table 5.2 presents environmental information sources mostly used by consultant respondents to draft EIS.

Five most used of information sources

Table 5.2 shows the five most used information sources answered by more than 50% respondents are sources related to technical information on environmental issues, environmental monitoring, environmental data, library and reports, and policies from central government. Meanwhile, none of the five most used information sources is related to flow of information. This suggests that the consultants used only information from their own office, possibly for practical consideration, or because sources from other offices were difficult to access.

Five most relevance of information sources

Less than 50% of total respondents indicated that the information supplied by all information sources as in the list was relevant to the EIS compilation. The moderately relevant information is technical information on environmental issues and environmental monitoring. This suggests that the consultants in conducting the environmental impact assessment were mostly focused to the technical aspects.

Five most needed of information sources

All information sources were identified by consultant respondents as moderate to less important information. Only one of the five most needed information, technical information on environmental issue, is regarded as very important by consultant, while the other four are regarded as moderately important information. The technical information on environmental issue is also was within the five most used and the five most relevant information sources to draft the EIS. This supports the statement that the consultants' view in conducting environmental impact assessment was mostly related to the technical assessment of the environmental impact.

Five most available of information sources

Table 5.2 shows that only the information related to courses and training of the five most available information sources is considered as being readily available, Furthermore, this information source is not within the five most used, five most relevant and five most needed information to draft the EIS. This suggests that the need for courses and training related to environmental impact assessment was saturated, and information generated from that activity was not considered important by consultant to draft the EIS.

Of the five most used information sources, three are part of the five most needed sources and part of the five most relevant information. These provide technical information on environmental issues, environmental monitoring, and local environmental data. However, the availability of these information sources was poor.

It is understood therefore, that the consultants themselves conducted the measurements accordingly.

Among the five most used information sources, the five most needed information sources and the five most available information sources, none consisted of information sources relating to the flow of information or networking. Flows of information between and within offices are reported by the consultants as "seldom used" because they are considered as "unnecessary" to draft the EIS. This shows that information sharing is considered unimportant by the consultants, and this would suggest that the consultants only used information sources available in their own office.

Table 5.2 presents information sources to draft EIS used by consultant respondents.

Table 5.2. Information sources mostly used by consultant respondents to draft EIS (n=17).

Sources of environmental information	**5 most used**	**5 most relevant**	**5 most needed**	**5 most available**
Sources for the EIA:				
Technical information on environmental issues	**65 %**	47 %	**71 %**	12 %
Status of natural resources		29 %	41 %	
Environment monitoring	**53 %**	41 %	47 %	
Practical advice from own professions			47 %	6 %
Interest and opinion of the community				6 %
Suggestions from experts				
Courses and training				**53 %**
Environmental data	**53 %**	29 %	47 %	6 %
Solutions to conflicts		29 %		6 %
Good experience from other areas				
Library and reports	**53 %**			
Policies from own office				
Policies from central government	47 %			6 %
Flow of information:				
Flow of information within own office				
Flow of information between offices				
Flow of information among professions				
From of information on related professions				6 %

Note: n: total respondents; %: % of total respondents that filled out option;
% in bold: % above 50%.

There are six main sources of information to draft the EIS defined in this research, namely, environmental monitoring data, environmental data, status of natural resources, technical information, flow of information between offices, and flow of information within an office. Figure 5.3 presents main sources of information used by consultant respondents, based on their perceptions on choices of "most used " and "not used." Their perceptions of choice "in between" the two choices of answers are not presented in this figure.

As in figure 5.3, the consultant respondents reported that the availability of all of the six main information sources to draft the EIS was poor. The availability of local data as monitoring data, environmental data, and status of natural resources is reported by more than 60% respondents as the most satisfactory. Although its availability was poor, most of the available local data was constantly used since they were highly needed. Four out of the six main information sources were highly needed. Two main information sources considered as moderate to the least needed wee related to flow of information or networking. This would suggest that the consultants used the available main information sources accordingly, only from their own office or from their own measurement data. Figure 5.3. shows main information sources used by the consultant respondents to draft EIS, based on their perceptions on choices of "most" and "not" used of information sources. Their perception on choice as "in between" is not represented in figure 5.3.

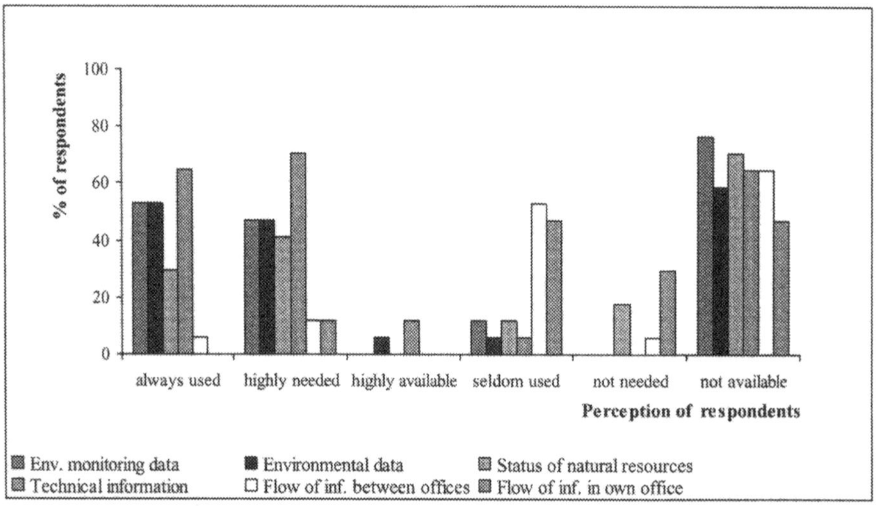

Figure 5.3. Main sources of information used by the consultant respondents to draft EIS (n = 17).

5.5. Information Sources Used by The Technical Teams and The District Governments to Review EIS

The sources of environmental information used by technical teams and district governments to review EIS are assessed on usage, how they are being perceived as necessary and relevant, and their availability. Each of these is assessed in three scales, "most," "in between" and "not." As an example, the usage is assessed as mostly used, used and not used. There are two groups of information sources, 13 sources related to EIA and 4 sources related to flow of information or networking. Discussions will first focus on the perception of technical teams and then of district government respondents on the five most used, five most needed, and five most available information sources. A summary of information sources mostly used by the technical teams and the district government respondents to review EIS are presented in tables 5.3 and 5.4.

5.5.1. Perceptions of the technical team respondents

Five most used of information sources

As table 5.3 shows the five most used information sources as selected by more than 50% technical team respondents were those related to technical information on environmental issues, environmental monitoring, environmental data, methodologies, and policies from central government. None of the sources related to flow of information was considered as important. This suggests that the technical teams only used information from their own office, as was reported by the consultant respondents. It was possibly for practical consideration, since they mostly used this information to review EIS. Unlike the consultants that considered library and reports as main important information to draft EIS, the technical teams had chosen information related to the methodology as one of the five most used information. This is understood, since the technical teams are responsible for guiding the conduct of the EIA and reviewing the quality of EIS, in particular by using methodologies for environmental impact assessment.

Five most relevance of information sources

None of the information sources was placed within the five most relevant information to review EIS, except technical information on environmental issues and monitoring. This suggests that the technical team respondents were particularly aware of the importance of technical information in environmental impact assessments.

Five most needed of information sources

Technical team respondents identified five information sources as moderate to very important to review EIS. Two of the five most needed information which were also within the five most used and the five most relevant information sources to review EIS, which were technical information on environmental issue and local environmental data were regarded as very important by technical teams. This would suggest that the technical teams view that in conducting the environment impact assessment the available environmental information sources needed to review EIS were appropriate.

Five most available of information sources

Table 5.3 shows that only one of the five most available information sources, the source related to courses and training was considered as moderately available. The other four information sources were reported as less available. Similar with the consultant reports, this information source was not within the five most used, the five most relevant and the five most needed information to review the EIS as reported by the technical teams. Of the five most used information sources, all were also in the five most needed sources and three were in the five most relevant information sources. These provide technical information on environmental issues, status of environmental resources, and good experience from other areas, in which its availability were poor. It is understood therefore, that the technical teams were commonly only review the available information in the EIS.

Among the five most used information sources as in table 5.3 and figure 5.4, the five most needed information sources and the five most available information sources, none consisted of information sources relating to the flow of information or networking. Flows of information between and within offices were reported by the technical teams as "less used" because they were considered as " less relevant" to review EIS. This shows that information sharing was considered unimportant by the technical teams to review EIS. This suggests that the technical teams relied on information sources available in their own office to review EIS, rarely using other information as a check and counter balance for reviewing information in EIS. Table 5.3 presents information sources to review EIS based on perception of the technical teams respondents.

Table 5.3. Information sources used by technical teams respondents to review EIS (n=28).

Sources of Environmental information	5 most used	5 most relevant	5 most needed	5 most available
Sources for the EIA:				
Technical information on environmental issues	**68 %**	39 %	**71 %**	11 %
Status of natural resources		36 %		
Environment monitoring	**61 %**		**54 %**	
Practical advice from own professions				
Interest and opinion of the community				
Suggestion from experts				
Courses and training				32 %
Environmental data	**64 %**	39 %	**71 %**	
Method of solution to the conflict	**61 %**	36 %	**50 %**	
Good experience from other areas				
Library and reports				
Policies of own office		25 %		14 %
Policies of central government	**64 %**		**57 %**	18 %
Flow of information:				
Flow of information within own office				11 %
Flow of information between offices				
Flow of information among professions		25 %		
Flow of information on related professions				

Note: n: total respondents; %: % of total respondents that filled out option;
% in bold: % above 50%.

As in figure 5.4, the technical team respondents reported that the availability of all of the six main information sources to review EIS was poor. Although availability was poor, most of the local data as monitoring data, environmental data, and status of natural resources was always used since they were necessary. Therefore, the technical team would only used limited sources of local information to review EIS, and they depended only on the information available in EIS to be reviewed. Although availability was poor, most of the local data as monitoring data, environmental data, and status of natural resources was always used since they were necessary. Two main information sources related to flow of information or networking was less to moderately used by the technical team. This would suggest that the technical teams seem to rely on data that is provided by the EIS. Figure 5.4 shows the main information sources used to review EIS by technical team respondents based on their perceptions on choices of "most" and "not" used of information sources. Their perception on choice as "in between" is not represented in figure 5.4.

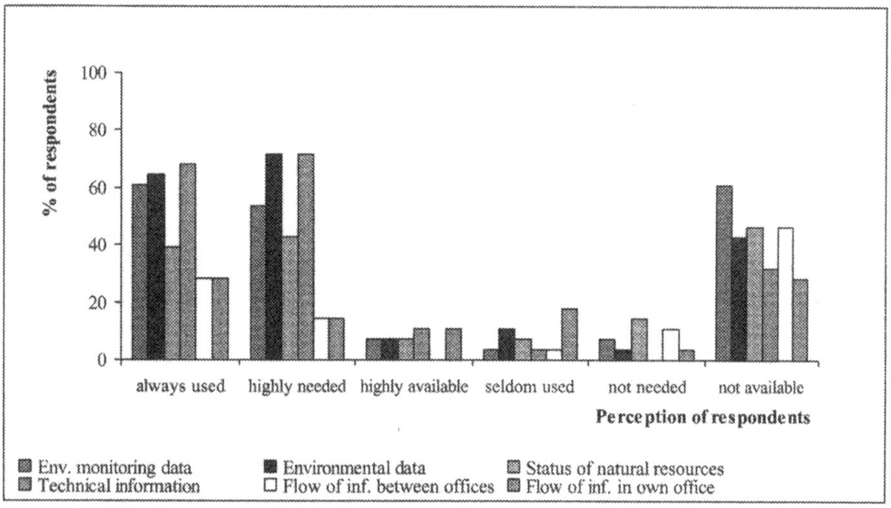

Figure 5.4. Main sources of information used by the technical team respondents to review EIS (n = 28).

5.5.2. Perceptions of the district government respondents

Five most used of information sources

Table 5.4 shows that of the five most used information sources as responded to by nearly 50% of total district government respondents were the sources of information related to courses and training and policies of the central government while the remaining sources were reported as being used less to moderately by the district government. One of these was related to flow of information within own office. This suggests that the district government only used limited sources of information to review EIS, especially information from their own offices.

Five most relevant of information sources

None of information sources as in the list falls within the five most relevant information to review EIS as stated by district government respondents. As in table 5.4, the five sources of information were reported as less to moderately relevant to review EIS. The moderately relevant information to the needs of the district government was information sources related to policies of the central government. This suggests that there is still a very close link between central government initiatives and district government responsibilities, despite the fact that the decentralisation of power to the district government has been introduced.

Five most needed of information sources

Two of the five most needed information, environmental monitoring and policies of central government were regarded as very important by district government respondents. Sources for EIA that consists of information on environmental monitoring and status of natural resources and local environmental data, which although moderately needed, its availability was low. None of information related to information flows was among the five most needed information sources. This suggests that the district government respondents were less aware of the importance of information exchange as a tool to gather the necessary information.

Five most available of information sources

Table 5.4 shows that only one of the five most available information sources, the source related to courses and training was considered by the district government as the only most available. The other four information sources are reported as less available. This information source was listed within the five most used information sources to review EIS as reported by the district government. This would support the assumption that there was a saturation of courses and training on EIA in Indonesia as reported by Crooks and Foley (1995). The district governments who are responsible for EIA data, however, did not have the data in practice. This support the argument stated in several publications that the district government was commonly lacking of information sources to conduct their tasks. As stated by Brown (1995) based on her research on the use of information sources by the Australian local governments, the local government respondents were not satisfied with the availability of the majority of the environmental information resources there.

Among the five most used information sources as in table 5.4, the five most needed information sources and the five most available information sources, none consisted of information sources relating to the flow of information or networking. Similar with the report of the technical team respondents, flows of information between and within offices were reported by the district government as "less used" because they were considered as " less relevant" to review EIS. This shows that information sharing was considered unimportant by the district government to review the EIS, and it would suggests that the district government mostly only used information sources available in its own office to review EIS. In contrast, Brown (1995) mentioned that flow of information between departments was mentioned in the first five information needs as stated by local government respondents in her research.

Table 5.4 present information sources to review EIS based on perception of the district government respondents.

Table 5.4. Information sources used by the district governments of Serang to review EIS (n = 14).

Sources of Environmental information	**5 most used**	**5 most relevant**	**5 most needed**	**5 most available**
Sources for the EIA:				
Technical info. on environmental issues		29 %	43 %	
Status of natural resources	29 %		36 %	
Environment monitoring			**57 %**	
Practical advice from own professions				
Interest and opinion of the community				
Suggestion from experts				
Courses and training	**50 %**		36 %	**64 %**
Local environmental data			43 %	
Method of solution to the conflict		29 %		
Good experience from other areas				
Library and reports				
Policies of own office	29 %	29 %	36 %	
Policies of central government	43 %	36 %	**50 %**	
Flow of information:				
Flow of information within own office	21 %			
Flow of information between offices		29 %		
Flow of information among professions				
Flow of information on related professions				

Note: n: total respondents; %: % of total respondents that filled out option; % in bold: % above 50%.

As in figure 5.5, the district government respondents reported that the availability of all of the six main information sources to review EIS was poor. Although its availability was poor, but most of the local data such as monitoring data, environmental data, and status of natural resources was always used since they were important. Therefore, the district government would only use limited sources of local information to review EIS. They only depend on to the available information contained in EIS to be reviewed. Two main information sources related to flow of information or networking was less to moderately used by the district government. This suggests that the district government seem to rely on data that was provided by EIS, and do not depend on other data from other offices for comparison in reviewing EIS.

Figures 5.5 shows the main information sources used by district government respondents to review EIS based on their perceptions on choices of "most" and "not" used of information sources. Their perception on choice as "in between" is not represented in figure 5.5.

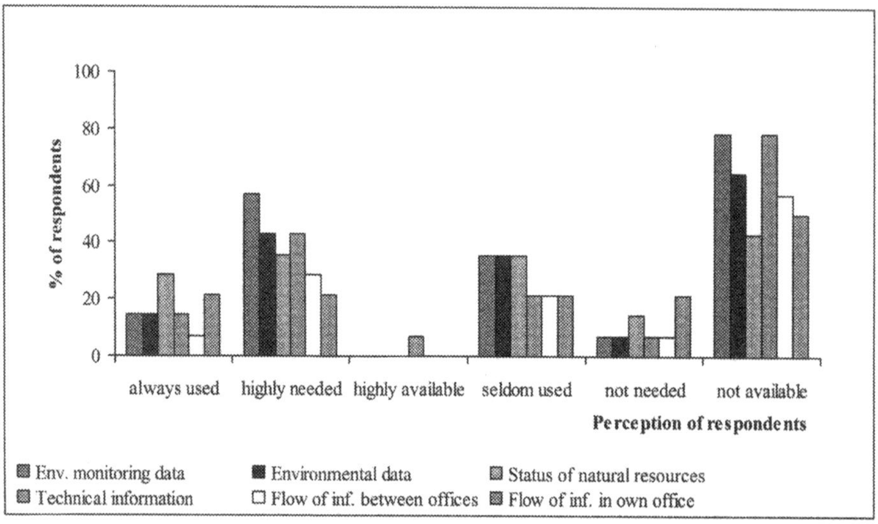

Figure 5.5. Main sources of information used by the district government of Serang to review EIS (n = 14).

5.6. Environmental Information Discussed in the Coastal EIS Review Meetings

Based on the method of observation of the coastal EIS review meetings described in section 4.4.5, the quality of environmental information discussed at the meetings has been evaluated from:

1. review on contents of the coastal EIS, and
2. discussions by participants in the coastal EIS review meetings.

5.6.1. Review of the coastal EIS contents in the meetings

One of the most difficult areas in review of EIS is ensuring objectivity of the review process, since the institution responsible for the review process may have a vested of interest in the decision making. There are various methods for promoting objectivity, among others through the use of review criteria. When the research was carried out, there were no standard criteria available to review contents of the coastal EIS at the meetings. Therefore, result of this research will be presented by EIS chapters: description of the plan, baseline conditions, significant impacts, evaluation of key impacts, mitigation plan, and monitoring plan.

Concerning the descriptions of the plan, only the Industry meetings utilised enough time to discuss the plan, the technical processes, and the schedule of activities of the proposed plan. The Sea Communications meetings omitted the discussion on this area as the participants, guided by the chairman, directly addressed key impacts and the proposed impact mitigation. They referred to the description of the plan only if it was necessary.

For the baseline conditions, all four meetings used enough time to discuss the physico-chemical condition of land and seawaters, but used limited time to discuss the status of land, biotic and socio-economic conditions, quantitatively and qualitatively. All meetings omitted the discussion on biological component. Discussions at the Sea Communications meetings were mostly on seawater quality, while the fishery component was slightly discussed. The marine biotic component was mostly ignored in the discussions. The reason given for this was that seawater was already polluted, the sediment was only mud, and no significant biotic component that needed to be conserved was present. The social component was also not discussed in detail, since the data were commonly taken from the local statistics office. The proximal islands were considered inhabited and the seawater surrounding was not considered a fishing ground of the local fishermen. This was a contradiction with what had been stated in a letter from the district government, received before the meetings, which stated that the area should be conserved from impacts caused by sea sand mining activities. The Industry meetings utilised enough time to discuss the baseline conditions, but only discussed the environmental components related to the technical aspects, such as the surface water and ground water quality. Limited discussion took place on the socio-economic aspects, and the land biotic aspects were slightly discussed, while the biotic aspects of the river and seawater environment were ignored at the Industry meetings.

All meetings concerning environmental impacts utilised enough time to discuss the key impacts of the plan to physical and chemical aspects of the environment, limited discussion was held on the socio-economic aspect while the biotic aspects of land and water were entirely omitted. Discussions at all meetings were mostly on the technical aspects of the key impacts to the main sectors of activity. The main discussions at the Sea Communications meetings were on key impacts of the plan to the environmental components related to the safety of sea transportation as the sector of main activities, while the impacts on other components were slightly discussed. The Industry meetings mostly discussed the key impacts to surface water and groundwater aspects, limited discussion took place on socio-economic aspects and biotic aspects were not discussed. In all meetings observed, enough time was utilised for discussions on assessment of key impacts of technical aspects. The time used to discuss linkages and interactions between the key impacts was limited.

All meetings concerning the impact mitigation plan used sufficient time to discuss technical details of impact mitigation. However, there was only limited discussion on details of mitigation measures and executing institutions, which are very important for the implementation of the mitigation plan. Most of the discussions at the meetings centred on measures to justify key impacts and did not really focus on how to prevent the impacts generated or even on the possibility to redesign the plan.

In contrast with time used for the discussion on the impact mitigation plan, only a very limited time was used in all meetings to discuss the components of the monitoring plan as well as the executing institutions. In all the meetings observed the linkage between mitigation plan and monitoring plan were never discussed. Limited discussion on the monitoring plan took place at the Sea Communication meetings. However in the Industry meetings the monitoring plan was not in the agenda. This part was considered almost to be out of the scope of the meetings compared to the mitigation plan. Table 5.5 presents a summary of observations of the coastal EIS review meetings.

5.6.2. Discussion by participants in the meetings

Generally, in most of the cases observed, there was an interval of less than 6 months between two meetings in each case (two Sea Communications meetings had a 3-month interval and one of the Industry meeting had a 4-month interval). One of the Industry meetings had a long time interval (1.5 years) because there were some constraints in land acquisition as the whole area mentioned in the EIS had not been purchased, and no permit had been issued by the district government on the use of water resources and electricity. In general all participants in the Sea Communications meetings got enough time before the meetings to review reports and to write inputs as the reports and invitations were sent one week before the meeting. For the Industry meetings, reports and invitations were distributed to the participants only three days before the meetings and usually it took some time for these to be delivered to the participants. Therefore they did not get enough time to review the reports and to write inputs.

In practice, less time than necessary was used for presentation and discussion of the coastal EIS in all meetings observed, approximately about 3-4 hours time was used in each meeting. Time used for the discussion was considered too short compared to the time used for the presentation of the contents of the EIS. The meetings should have had more time allocated for discussions aimed at getting inputs for improvement of the quality of the EIS. For the Industry meetings, the discussion on administration documents such as permits, letters of recommendation, and supporting data was considered enough. The Sea Communications meetings concentrated on discussions of technical issues. The chairman of the latter meeting has said that the administrative issues should be resolved outside the meetings between the initiator and the institution granting permits. There were inconsistencies between the documents and their presentations in all four meetings. Generally all participants consistently followed most of the discussions with guidance from the chairman of each meeting and provided some comments and input.

In the case of Sea Communications, the consultants held discussions with the technical personnel before the meetings and made revisions to the documentation. Therefore the image created at these meetings was that participants mostly agreed with the contents of the documents. On the other hand, at the Industry meetings, the consultants appeared to consider the quality of the documentation as low and there was criticism of the consultants by the other participants.

Table 5.5 presents a summary of observations of the 4 coastal EIS review meetings.

Table 5.5. Summary of observations of the 4 coastal EIS review meetings.

	Industry sector		Sea Communications sector	
Review criteria	**Meeting 1**	**Meeting 2**	**Meeting 3**	**Meeting 4**
A. *Time utilisation*				
1. Preparation time for participants				
a. To review the EIA documents	L	L	E	E
b. To write comments	L	L	E	E
c. To prepare the presentation (for consultant)	E	E	E	E
2. Implementation of the meeting				
a. Presentation	E	E	E	E
b. Discussion	E	E	E	E
B. *Presentation of document*				
1. Administrative completeness				
a. Permits	E	E	L	L
b. Letter of recommendations	E	E	L	L
c. Supporting data (laboratory results, etc.)	E	E	L	L
d. Tables/Maps/graphics	E	E	E	E
2. Environmental information				
a. Discussion in the meeting	L	L	L	L
b. Discussion in the decision making	L	L	L	L
3. Consistence				
a. Between tables and written explanation	L	L	E	E
b. Between document and presentation	L	L	E	E
c. Between stages within presentation	L	L	E	E
C. Discussion of the EIS contents				
1. Description of the Plan				
a. Stages of activity within the Plan	E	E	L	L
b. Detail Plan/ Technical process	E	E	L	L
c. Detail schedule of activity	E	E	E	E
2. Baseline conditions				
a. Physical/Chemical (land/sea waters)	E	E	E	E
b. Land, biotic	L	L	L	L
c. Waters, biotic (river and sea)	N	N	N	N
d. Socio-economic and Culture	L	L	L	L
3. Key impacts				
a. Physical/Chemical (land/sea waters)	E	E	E	E
b. Land, biotic	N	N	N	N
c. Waters, biotic (river and sea)	N	N	N	N
d. Socio-economic and culture	L	L	L	L
4. Evaluation of key impacts				
a. Linkages between key impacts	L	L	L	L
b. Interaction between key impacts	L	L	L	L
c. Assessment of key impacts	E	E	E	E
5. Mitigation measures plan				
a. Linkages with key impacts evaluation	L	L	L	L
b. Technical details/ mitigation measures	E	E	E	E
c. Detail of locations for mitigation measures	L	L	L	L
d. Schedule for mitigation measures	L	L	L	L
e. Implementing and advisory institution	L	L	L	L
6. Monitoring Plan				
a. Linkage with the Mitigation measures plan	N	N	N	N
b. Monitoring parameters	N	N	L	L
c. Detail of monitoring methods	L	L	L	L
d. Detail of monitoring locations & schedules	L	L	L	L
e. Implementing and advisory institution	L	L	L	L

Note: E = Enough time, L = Limited time, N = No time.

5.7. Monitoring of Environmental Parameters

The use of environmental information for environmental monitoring is assessed based on in-depth interviews with the district government respondents on how parameters of environmental components and environmental impact components were being monitored. The results are divided into two parts:

1. parameters for environmental monitoring as well as constraints in the monitoring, and
2. parameters for environmental impact monitoring and constraints in the monitoring.

5.7.1. Parameters for environmental monitoring

Parameters for environmental monitoring consist of six groups of components: hydrology and river water quality; hydro-oceanography; spatial and land use plans; flora; fauna; and socio-economic-culture. Table 5.6 shows perceptions of respondents on parameters used for environmental monitoring, and table 5.7 shows constraints in using parameters for environmental monitoring.

- In general, the respondents reported that most parameters on hydrology and river water quality were less often used for daily environmental monitoring. Parameters often used for environmental monitoring were related only to river water quality, such as temperature and levels of Biochemical Oxygen Demand (BOD) and Chemical Oxygen Demand (COD) in the river water. In the implementation of monitoring, they reported some constraints faced that were more related to lack of laboratory facilities, limited human resources, and financing.
- Most of parameters within hydro-oceanography components were also reported by respondents as less often used for environmental monitoring. Parameters reported as often used for monitoring were only related to seawater quality and include the temperature, level of BOD-COD, mineral content, colour and turbidity. They were similar to parameters often used in river water monitoring.
- Constraints that hindered the measurement of indicators by the respondents were mostly related to laboratory facilities, human resources, data recording and analysis, and financing. This was commonly experienced by district government in monitoring parameters related to hydro-oceanography components as reported by the respondents.
- Parameters related to spatial and land use plans were reported as less often used for environmental monitoring by the respondents. The constraints experienced in the implementation of monitoring for these parameters were reported to be mostly in recording, analysis, and financing.
- Parameters on land and river flora were less often monitored, while parameters on flora and habitat related to beach and marine were never been monitored. Parameters within the biological component that were related to fauna were reported as never used for environmental monitoring. Constraints experienced were reported in human resources, data recording, analysis, and financing.

- Half of the parameters within the socio-economic-culture component were reported to be often being monitored by the operators, while another half was reported to be less often monitored. For those parameters that were reported as often being monitored were commonly being monitored by other offices, as the District Statistic Office. For those parameters that were less often used for monitoring, the reported constraints were related to data recording and analysis.

Table 5.6. shows environmental monitoring as indicated by the district government of Serang and table 5.7 shows constraints in environmental monitoring.

Table 5.6. Environmental monitoring indicated by the district government of Serang.

	Frequency of monitoring		
Environmental Component	**Often**	**Less often**	**Never**
1. ***Hydrology and river water quality***			
1.1 Physical characteristics of river and swamp		●	
1.2 Average debit for decade, month and year		●	
1.3 Sedimentation and erosion rates		●	
1.4 Water quality : physical, chemical and microbiology	●		
2. ***Hydro-oceanography***			
2.1 Hydrodynamics pattern		●	
2.2 Sedimentation or abrasion		●	
2.3 Sea water quality: physical, chemical and Microbiological	●		
2.4 Physical oceanography		●	
3. ***Spatial and land use***			
3.1 Land use		●	
3.2 Spatial plan, Land Use plan		●	
3.3 Possibility of changes of Spatial and Land Use plans		●	
3.4 Number and condition of historical buildings		●	
3.5 Aesthetically value and landscape scenery		●	
3.6 Infrastructure for public transport and cargo		●	
3.7 Clean water infrastructure		●	
3.8 Waste water infrastructure		●	
3.9 Electricity infrastructure		●	
4. ***Flora***			
4.1 Land flora and habitat		●	
4.2 River flora and habitat		●	
4.3 Beach and marine flora and habitat			●
5. ***Fauna***			
5.1 Land fauna in general			●
5.2 Land insects (important insects)			●
5.3 Water fauna and habitat			●
5.4 Beach and marine fauna and habitat			●
5.5 Important land fauna			●
5.6 Important marine and fresh water fauna			●
6. ***Socio-economic-culture***			
6.1 Existing socio-economy and culture	●		
6.2 Settlement and employment	●		
6.3 Community's perception of the plan		●	
6.4 Interaction with community before and after the project		●	

Table 5.7. Constraints in environmental monitoring.

Environmental components	Constraints in environmental monitoring			
	Laboratory facilities	**Human resources**	**Recording & Analysis**	**Financing**
1. Hydrology and river water quality	•	•		•
2. Hydro-oceanography	•	•	•	
3. Spatial and land use		•	•	•
4. Flora		•	•	
5. Fauna		•	•	•
6. Socio-economic-culture			•	•

5.7.2. Parameters for environmental impact monitoring

Parameters for environmental impact monitoring are grouped into seven components: changes of physiography; changes of hydrology and surface water quality; changes of hydro-oceanography; changes of spatial plans; changes of land and water flora; changes of land and water fauna; and changes of socio-economy and socio-culture. Parameters for environmental impact monitoring are reported by district government respondents as follows:

- In general, parameters related to changes of physiography component were reported by the respondents as less often or never used for environmental impact monitoring, the reason being that they had limited human resources for recording and monitoring these parameters.
- Parameters on the changes of hydrology and river water quality were also reported as less often used for impact monitoring. Constraints in the implementation of these parameters were related to limited human resources and limited skills in recording and analysis. The only parameters reported as often used for environmental impact monitoring were those related to changes of surface water quality and ground water quality such as temperature, level of BOD-COD, mineral content, colour, and turbidity.
- Most of the parameters within the changes of hydro-oceanography component were reported as less often being used for monitoring environmental impacts. Parameters often used for impact monitoring were related to seawater quality, such as temperature, level of BOD-COD, mineral content, colour, and turbidity. Constraints for using other parameters for impact monitoring were reported to be mostly limited to laboratory facilities, lack of skills for data recording and analysis, and financing.
- Parameters related to changes in spatial and land-use plans for environmental impact monitoring were reported as less often being monitored. The reported constraints related to monitoring of changes in the land use plan and constraints in recording and analysis.
- There was no reasons given for not monitoring other parameters related to infrastructure, and this might be because other offices had monitored these parameters.
- The respondents reported that the parameters related to the changes in land and water flora and fauna were less often monitored in all case studies. They also reported that the biological component was less often used for environmental

impact monitoring purposes. The constraints here were related to human resources, limited skills in recording and analysis of data, and financing.

- Almost all parameters under the socio-economy-culture component were reported by the respondents as less often or never used for impact monitoring purposes. Possibly, because other office as the District Statistic Office was responsible in monitoring this component. Constraints reported were mostly related to data recording and analysis.

Table 5.8 shows perceptions of the district government of Serang on environmental impact monitoring, and table 5.9 shows constraints in environmental impact monitoring.

Table 5.8. Environmental impact monitoring indicated by the district government of Serang.

Affected environmental components	Frequency of monitoring		
	Often	Less often	Never
1. *Changes in physiography*			
1.1 Slope stability		●	
1.2 Land erosion		●	
1.3 Changes/modification of land			●
2. *Changes of hydrology and surface water quality*			
2.1 River flow		●	
2.2 Water surface and water body		●	
2.3 Flow direction and river flow pattern		●	
2.4 Flood magnitude and extent of damage			●
2.5 Formation of water stagnation		●	
2.6 Water quality : physical, chemical, microbiological characteristics	●		
2.7 Groundwater quality	●		
3. *Changes of hydro-oceanography*			
3.1 Hydrodynamics pattern		●	
3.2 Sedimentation or abrasion patterns		●	
3.3 Sea water quality: physical, chemical, and microbiology	●		
3.4 Process of physical oceanography		●	
4. *Changes of spatial plans*			
4.1 Land use and other natural resources	●		
4.2 Scenery		●	
4.3 Historical and conserved buildings		●	
4.4 Suitability to the regional planning		●	
4.5 Pattern changes in public transportation		●	
4.6 Clean water and electricity services		●	
5. *Changes of land and water flora*			
5.1 Loss and damages to the general plant community		●	
5.2 Loss and damages to the rare plants		●	
5.3 Changes, loss and damages to highly economic plants		●	
6. *Changes of land and water fauna*			
6.1 Migration pattern (fish, reptile, mammal, plankton)			●
6.2 Breeding and animal territory areas for fish, bird, etc.		●	
6.3 Mortality of fish and birds due to environmental pollution		●	
6.4 Animal extinction, or scarce fish or highly economic fish		●	
6.5 Habitat disturbance		●	
7. *Changes of socio-economy and socio-culture*			
7.1 Existing economy, tourism, trade, infrastructure, and employment		●	
7.2 Population and employment structure	●		
7.3 Attitudes, values and perceptions of groups of people		●	
7.4 Social stratification system		●	
7.5 Existing local institutions	●		
7.6 Condition of power distribution system		●	
7.7 Damage to historical and conserved buildings		●	
7.8 Land speculation		●	
7.9 Public health facilities and public safety		●	

Table 5.9. Constraints in environmental impact monitoring.

Affected environmental components	*Constraints in environmental impact monitoring*			
	Laboratory facilities	**Human resources**	**Recording & analysis**	**Financing**
1. Changes of physiography		•	•	•
2. Changes of hydrology and surface water quality	•	•	•	•
3. Changes of hydro-oceanography	•	•	•	•
4. Changes of spatial plan	•		•	
5. Changes of land and waters flora		•	•	
6. Changes of land and waters fauna		•	•	
7. Changes of socio-economy and culture		•	•	•

5.8. Environmental Information Based on Perceptions of Technical Teams

The technical team is continuously involved in most stages of EIA process. Therefore, its perception of the use of environmental information in the EIA process is very relevant for the analysis of this research. Perceptions of technical team respondents in regard to the EIA process consist of perceptions on:

1. use of the EIA technical guidelines,
2. environmental information in the EIA guidelines and aspects in EIS,
3. prospects of the EIA due to the implementation of the regional autonomy law, and
4. suggestions for improving the EIA in the coastal zone.

5.8.1. Use of the EIA technical guidelines

Respondents have been asked about the use of the EIA technical guidelines. In general, all respondents reported that the technical guidelines were always used to review the EIS of the projects within their sector of activity. However, respondents gave answers ranging from “always”, “seldom” to “never” concerning the use of these guidelines in the post-EIS stage, the stage after the EIS is approved, and when impact monitoring is being conducted. A possible reason for this might be that their answers would refer to the intended use of the guidelines as stated in the law, and they were only dealing with the EIS review and did not have activities related to the monitoring. Similarly, Donelly *et.al.* (1998) in their study showed that the use of the guidelines was avoided at the earliest stages as well as at the later stages of the impact assessment process, especially in implementing mitigation and monitoring plan. The EIA guidelines seem only being used for impact assessment purposes. Similar conclusions had been recorded by Sadler and Verheem (1996) based on their study of the effectiveness study of country status in the EIA.

Respondents were asked to indicate anticipated benefits of the use of the EIA technical guideline and benefits achieved in practice. The benefits of the EIA guideline anticipated by respondents are:

- to clarify the EIA study,
- to clarify the reporting and reviewing,
- to simplify procedure and methodologies,
- to clarify impact assessment,
- to simplify decision-making process,
- to simplify the implementation, and
- as a control tool in making and reviewing the EIS.

Although the respondents expected all these benefits of the use of the technical guideline, in practice all the benefits were not achieved. Among the benefits reported as not achieved was simplification of the decision-making process, implementation of mitigation and monitoring plans, and use of the guidelines as a control tool. There can be several reasons behind this reported failure to achieve benefits. The processes of planning and EIS compilation were separate and were conducted by different consultants, while sharing of data and information between the consultants was limited. The permit approval process for planned development was delayed in some cases since several offices granted permits and there was no co-ordination between the permit systems. Furthermore, there was possible political influence in the national sector offices, which grant permits and also approve the EIS, which in some cases hindering the function of the technical guideline as a control tool in the EIA process. The technical guideline was being used only for guiding the EIS compilation, and does not refer yet to the bigger aim of the technical guidelines, which is environmental conservation. Almost all respondents reported that the technical guidelines of their sector were too general, and were not presented in simple language. One of the respondents added that the present guidelines were like static regulations, since they were never updated and they did not cope with problems experienced in the field. Donelly *et.al.* (1998), based their study on the review of EIA guidelines report a similar trend and highlight a number of key deficiencies in the quality of guidelines. With few exceptions, most guidelines seem static documents as they do not seem to be revised through regular updating to keep pace with changing ideas on best practice. Furthermore, Spooner (1998) highlighted factors constraining the use of guidelines due to lack of legal status; limited time and financial resources for the EIA process; lack of institutional co-ordination; lack of awareness of the contents as the contents are often too technical and also that the guidelines often over emphasise negative impacts while under emphasising its positive aspects.

5.8.2. Environmental information in the EIA technical guidelines

Concerning environmental information in the technical guidelines, respondents from the Sea Communications and the Industry sectors agreed that only the socio-economic component in their guidelines had parameters with sufficient indicators. They evaluated other components differently; in which components related to their sector of activities had sufficient parameters. The respondents from the Sea Communications sector mentioned that only the parameters related to the marine component in their guidelines had sufficient indicators. The respondents of the Industry sector assessed that the hydrology and river water components in their technical guidelines had sufficient indicators. Most respondents viewed parameters within their own sector as having sufficient indicators, while parameters outside the activity of their sectors were viewed as having insufficient indicators. Possible explanation of this might be because experts from a sector of activity influenced the development of technical guideline of that sector. Regarding the weaknesses of the technical guideline, respondents reported that the guidelines of their sector of activity were too general, and needed to be revised regularly every 3-5 years, to be able to anticipate problems that keep changing.

The respondents were asked to review important legal aspects that should be described in EIS. They selected 8 out of 11 important aspects listed and also added impact mitigation, monitoring plans and sustainability of the plan to this list. The aspects selected as unimportant were alternatives for design review, assessment of proposed alternatives, and alternatives for providing compensation. As the respondents had not considered design review as an important component of the EIA, it seems that they viewed the EIA process as an attachment or addition to the design plan or to the feasibility study of the plan, and not as part of the design plan. The EIA was seen as a justification to the impact generated by the plan. Ideally, the EIA should be part of an overall design plan, so if there is a possibility that the plan will generate a significant negative impact on the environment, it can be prevented by some adjustment in the design plan. Thus it is seen that the objectives of the EIA were not properly understood.

5.8.3. Implementation of law on regional autonomy and its effect on the EIA process in 1999

The most significant progress of the EIA process in 1999, as mentioned by the respondents, was the inclusion of stakeholder participation and access to environmental information in the EIA process with the introduction of decree no. 08/2000. The direction of EIA activities, according to some respondents, was in some cases increasingly moving away from the intended objective of the EIA, which was to conserve the environment from impacts of the proposed plans. There was no available guideline to review EIS, or monitor activities in the implementation of EIA decisions. The only available guideline was a procedure to draft EIS.

Concerning decentralisation, there was a general agreement among the respondents that this was a significant improvement toward the success of EIA processes. It was also viewed as a way to improve stakeholder participation in development activities as well as to conserve the environment. The district stakeholders, who knew their area better, will have better access to the information on the development plans in their area. They will be aware of possible impacts generated by the development plans and the impact mitigation and monitoring measures. Therefore they will be responsible for their decisions, especially those related to the sustainability of their environment. The respondents also mentioned possible changes to the existing law no. 20/1999 on regional autonomy and law no. 27/1999 on the handing over of EIA approval to the district levels. The passing of the law on regional autonomy implied changes in responsibilities to review and to approve EIS from central government to district government.

The respondents agreed that the progress of decentralisation of the EIA needs a lot of preparation concerning institutional arrangements as well as human resources, both at central and local levels. Furthermore, they indicated some weaknesses in reviewing and approving EIS, such as limited capacity of local human resources and institutions in the EIA process within the district government. They agreed that training of human resources within district governments should be considered as a priority, especially in gaining knowledge of the technical aspects of environmental impact assessment. The respondents from the Sea Communications sector specifically mentioned the importance of regulations on sizes of projects in the EIA.

5.9. Conclusions

Research question 1: How is environmental information stated in the EIA guidelines?

According to the reviewers, the Indonesian guideline had enough indicators compared to the two technical guidelines. Furthermore, the technical guidelines should have more indicators. The sufficiency of indicators of parameters within the general guideline and the technical guidelines should be similar. Whatever the focus of the sector, the technical guidelines should have sufficient indicators to describe the quality of the parameters. The respondents viewed parameters within their own sectors as having sufficient indicators while those not having relations with their sector of activities were viewed as having insufficient indicators. A possible reason for this opinion was that the developer of the guidelines does not know the importance of other parameters that have no direct relation to their sectors.

As seen in table 5.1, the environmental information on the biological component was lacking since all guidelines reviewed were reported to have insufficient indicators within the biological component. Similarly, respondents of the technical team (section 5.8.2) reported that although the technical guideline containing sufficient environmental information related to their sector activity, the information contained was general because information that was not directly related to their sector was limited. This could be due to the fact that the developer of the guidelines although having sufficient knowledge in the technical aspect lacks the knowledge of the biological component. The insufficiency of knowledge on the biological component, especially marine biology would indicate a limited understanding of the importance of the biological component in the environmental impact assessment in Indonesia. A possible explanation for this maybe due to experts on one sector of activity influenced only the development of their technical guideline while the support from other experts not connected to the project was less. Moreover, the technical guidelines were developed in the early 1990s during which period marine research was still limited in its accessibility, despite the intensive co-operation research programmes conducted in Indonesia as reported by Dahuri and Dutton (2000).

All guidelines reviewed have not mentioned executing institutions and related institution for impact monitoring. The Industry technical guideline and the Indonesian EIA general guideline mentioned an institution only for impact mitigation while the Sea Communications guideline did not name any institution for impact mitigation and monitoring. Legally no institution has the executing authority for impact monitoring. Therefore, monitoring of the EIA decisions has not been done. This shows that mitigation activities are considered more important than monitoring activities in Indonesia due to lack of awareness of impact monitoring. The new law on Regional Autonomy gives the district government more power concerning environmental conservation, in particular the establishment of local institutions for impact monitoring.

Research question 2: What is the quality of environmental information described in the coastal EIS?

Based on the review of the coastal EISs and the application of total weighted-ranks for sensitivity analysis, the coastal EISs qualities for each sector and all reports can be summarised as follows. Although approved by the authority, 5 out of 10 reports have a quality above the approval level, and another 5 below the approval level. All reports on average have a quality below the approval level, as seen in figures 5.1 and 5.2. The Sea Communication reports on average have sufficient quality for the approval. However, the Industry reports on average do not have satisfactory quality for approval. Average scores of the Sea Communications reports on all review areas are higher compared to the average scores of the Industry reports. The average scores of the Sea Communications reports in the review area on general description and impact mitigation and monitoring measures are higher (B) compared to other review areas (C). Similarly, the average score for the Industry reports in the review area on general description (1) is higher (D) compared to other areas (E). Similar results are identified for EIS quality studies in developing countries. Most reports (70%) of Indian case studies are unsatisfactory and more than half of reports for Tanzanian case studies are below the acceptance level (Mwalyosi *et.al.,* 1998). Lee (2000) furthermore reported that in both developing and developed countries, there is an indication of EISs of poor quality being produced.

In this study, average scores in their overall assessment for each sector are higher for the dominant cases compared to the scores obtained by all reports. It means that the reports for the dominant cases have better quality. An explanation of the better quality in the four dominant cases in this research could be that they have a potential to create an environmental impact and hold public attention. Since their projects are big projects, they have more money to spend on better consultants, on better field surveys and analysis as well as on the EIS compilation. Furthermore, the highly public attention to these dominant cases ensures that the work of technical teams in assisting the EIS compilation and reviewing the EIS are done thoroughly. The technical teams from related sectors put in their best efforts in reviewing the EIS, in order to avoid public criticism if there is pollution in the future. Therefore the developers have tried to improve the EIS to show very clearly how to minimise potential environmental impacts generated in each case.

Based on the review of indicators in the technical guidelines, it is observed that the guideline for the Industry sector is more sufficient in indicators than the guideline for the Sea Communications sector. Although the Industry sector has a better technical guideline, the average quality of the EISs is low. But, in the Sea Communications sector, the EISs average quality is high although the technical guideline is weaker. Thus the quality of the EIS shows a contradiction with the quality of the technical guideline per sector, which indicates that the quality of the EIS is influenced more by the way the EIS is composed and reviewed. Possible explanation for this indication among others is cases of the Sea Communications sector have less variety compared to the Industry sector, so the EIS review is conducted more thoroughly in the Sea Communications sector.

The general observations on the coastal EISs reviewed in this study indicated that they are weak in the following areas:

- assessment areas of description on baseline condition,
- scooping of impacts,
- assessment of significance impacts,
- commitment to monitoring measures, and
- communication of results including uncertainties as well as non technical summary.

Similar findings are reported by Ibrahim (1992) based on his research of EISs on Malaysian projects and Mwalyosi *et.al.* (1998) based on their research of EISs on Tanzanian and African projects. Biswas (1992) reported similar trend experienced by developing countries adding that these EIA reports are mostly too mechanistic, concentrating mostly on the technical aspects of impact assessment, while any discussion if any tends to be somewhat general and qualitative.

Research question 3a: What environmental information sources are used by consultants to draft EIS?

In general, the consultants were not supported by sufficient sources of important information to draft EIS. This is seen in table 5.2 and figure 5.3. None of information sources was found within the five most used, the five most needed and the five most relevant sources of information to draft EIS. Two of the five most used of information sources were found as moderate to highly needed information, and no information was considered among the five most relevance information. The six main information sources as seen in figure 5.3 that are important to draft the EIS were moderately unavailable and four sources fall within the least available group, according to the consultants. Most of the important local data, which include the status of natural resources, environmental data and environmental monitoring data, although highly needed, were reported to be relatively unavailable. This indicates that sources of local environmental information needed by consultants to draft the EIS were scarce. Information related to courses and training was considered as sufficient in availability. But this source was not within the five most used, five most relevant and five most needed information to draft the EIS. This suggests that courses and training related to environmental impact assessment had been saturated since 1989 as reported by

Crooks and Foley (1995), and information generated from these activities were not considered important by consultant to draft EIS. Among the five most used information sources, the five most needed information sources and the five most available information sources, none consisted of information sources relating to the flow of information or networking. Flows of information between and within offices were reported by the consultants as "seldom used" because they were considered as "unnecessary" to draft EIS. This shows that information sharing was considered unimportant by the consultants. This means that the channel of information to obtain data from other sources, which is important to draft the EIS, was not properly open and sources from other offices were difficult to access. This is a possible explanation why the information sources used to draft the EIS were mainly those within the consultant own office.

Research question 3b: What environmental information sources are used by the technical teams and district governments to review EIS?

The important information sources related to the EIA data (methodology, local environment data and policies) were reported by most respondents of technical teams and district governments to be "moderately unavailable." Although the important information sources to review the EIS were not sufficiently available, the technical teams were supported by more important sources of information to review EIS, compared to the district government. Of the five most used information sources as reported by technical teams, all were found in the five most needed, three were in the five most relevant information sources, and none was found in the five most available information sources. The district government considered only one of the five most available sources of information as being moderately accessible to review EIS. The other four information sources were reported to be less accessible. This source was listed within the five most used information sources to review EIS as reported by the district government. This supports the assumption that courses and training on EIA was saturated in Indonesia as reported by Crooks and Foley (1995). The district government, which is responsible for local environmental data, was not supported by sufficient information sources to review EIS. One reason for this was found in the unavailability of flow of information sources. Respondents considered flow of information between offices as unimportant (see tables 5.3 and 5.4, figures 5.4 and 5.5). Sharing information, which is important to obtain data from other offices, had not been successful within and between groups. This explains the insufficiency of data used to review EIS as experienced by the technical teams and the district government respondents. This supports a point raised in several publications that the district government lacks information sources to conduct their tasks. A national survey on information needs of local governments for environmental management in Australia shows similar results (Brown, 1994). The study found that 28 types of information were relatively unavailable and 5 of the most important types of information were the least available. The local government respondents were not satisfied with the availability of the majority of the environmental information resources there. The study also showed that local government was experiencing constraints in managing their areas due to closed communication circles within and between offices. The district governments who are responsible for EIA data, however, did not have the data in practice. In the Indonesian system where the national sector offices dominate the local level, environmental information is recognised as relatively unimportant compared to information related to a particular sector of activities. For example, in the Sea Communications sector, only local data related to its main activity-safety of marine transportation was collected at national level. Unfortunately, the data were mostly categorised as confidential and was available only within its offices. Other data, although part of marine data, were not collected since they did not have any direct relation to marine transportation.

Research question 4: How is the use of environmental information in the coastal EIS review meetings?

In general, although environmental information directly related to the sector main activity was sufficiently discussed at all EIS review meetings although information not directly related was not. The physical condition of the environment was sufficiently discussed, information related to biology and social-economic components was not sufficiently discussed. For the baseline conditions, all four meetings used enough time to discuss the physico-chemical condition of land and seawaters, but used limited time to discuss the status of land and socio-economic conditions. Sufficient discussion on the basic condition and on the key impacts of the plan related to the physico-chemical conditions of land and seawaters, that related to the sector main activity, took place at all meetings while limited discussion took place on land biotic and socio-economic conditions. Discussions on the baseline condition related to the marine biology component were ignored at all meetings. The social component was also not discussed in detail, since the data were commonly taken from the local statistics office. The proximal islands were considered inhabited and the seawater surrounding was not considered a fishing ground of the local fishermen. The technical aspects of the impacts on the environmental component of the development plan which related to the sector main activity were thoroughly discussed at all meetings. But discussions on impact mitigation measures were given priority over impact monitoring measures. The discussions concentrated mostly on justification of the measures to mitigate the environmental impacts generated. Executing institutions for impact mitigation were also discussed. Impact monitoring was considered not important at all meetings, since a limited time was used in all meetings to discuss all the important aspects of the monitoring measures and executing institutions for monitoring was not a topic for discussion.

Research question 5: What environmental parameters and indicators are monitored by the district government?

Most of the parameters on environmental monitoring were reported as less often or never monitored by the district government respondents, and most of the expressed constraints to monitoring were related to human resources, data recording and analysis, and funding. The same problems were found for parameters within the environmental impact components, where the respondents stated that most of the parameters within these components were less often or never monitored. Parameters often used for environmental monitoring were related to river water and sea water quality, such as BOD, COD, mineral content, colour and turbidity. Parameters under socio–economy component were often monitored by the District Statistic office.

The EIA guidelines did not state which department of the district government will be responsible for environmental impact monitoring. Therefore the identification of the executing institution was one of the problems encountered in the monitoring of implementation of the post-EIA stage. Another problem was law enforcement. Here too, enforcement was limited since the monitoring institution was not yet clearly authorised. The developers or the initiators did not take seriously the penalties stated in the local regulation.

The lack of environmental monitoring conducted by the district government as experienced in this case study is also recognised phenomenon in other developing countries as well. Wood (1995) for example, surmised that monitoring had been a missing step in the EIA process. Similarly, Ahmad and Sammy (1992) based on their study of EIA in developing countries indicated that there was very little EIA monitoring conducted. Biswas (1992) had noted that environmental controls may not be monitored due to the fact that projects may change substantially between EIS approval and its implementation. Furthermore, Dahuri and Dutton (2000) remarked that lack of monitoring as experienced by the local government on coastal management among others was due to limited skills and knowledge on coastal and marine resources and processes.

Research question 6: How is environmental information used in the EIA process based on perceptions of the technical teams?

The technical teams reported that the technical guidelines were always used to review the EIS, however they reported as "always", "seldom" to "never" concerning the use of these guidelines in the post-EIS stage, the stage after the EIS is approved, and when impact monitoring is being conducted. Sadler and Verheem (1996) and Donelly *et.al.* (1998) reported similar conclusions that the use of the guidelines was avoided at the earliest stages as well as at the later stages of the impact assessment process, especially in implementing mitigation and monitoring plan.

Concerning environmental information in the technical guidelines, respondents from the Sea Communications and the Industry sectors agreed that only the socio-economic component in their guidelines had parameters with sufficient indicators. Most respondents viewed parameters within their own sector as having sufficient indicators, while parameters outside the activity of their sectors were viewed as having insufficient indicators. Possible explanation of this might be because experts from a sector of activity influenced the development of technical guideline of that sector.

Concerning decentralisation, there was a general agreement among the respondents that this was a significant improvement toward the success of EIA processes. It was also viewed as a way to improve stakeholder participation in development activities as well as to conserve the environment. The respondents also mentioned the passing of the law on regional autonomy implied changes in responsibilities to review and to approve EIS from central government to district government. The respondents agreed that the progress of decentralisation of the EIA needs a lot of preparation concerning institutional arrangements as well as human resources, both at central and local levels. Furthermore, they indicated some weaknesses in reviewing and approving the EIS, such as limited capacity of local human resources and institutions in the EIA process within the district government.

6. STAKEHOLDER PARTICIPATION IN THE COASTAL EIA PROCESS

6.1. Introduction

This chapter presents an analysis of stakeholder participation in the coastal EIA process based on the following assessments:

- perceptions of technical teams (section 6.2),
- stakeholder participation in the coastal EIS review meetings (section 6.3), and
- stakeholder participation in the pre-coastal EIS compilation (section 6.4).

6.2. Perceptions of Technical Teams on Stakeholder Participation

The technical teams' perceptions on stakeholder participation in the EIA process, based on government regulation no. PP 51/1993 and on new regulation no. 08/2000 on stakeholder participation, are considered in this research. During the period under consideration, the national-level sectors of government had authority to process and approve the EIS review. The Ministerial Decree for Environment no. 40/2000 on the delegation of EIS approval to the local level and decree no. 41/2000 on the formation of the EIS review committee at district or city level were not yet implemented when this research was carried out (1999-2000). Discussions will first focus to the decrees on stakeholder participation, and later on the perceptions of technical teams on stakeholder participation in the EIA process.

The decree no. PP 51/1993 on the EIA process does not clearly address guidance on stakeholder participation, and it only advocates the initiators or the developers to only inform the public concerning the EIA process. It does not define the specific levels of society who should be informed and the level of public involvement in the EIA process. The new decree no. 08/2000 is more advanced compare to the decree no. PP 51/1993. It defines the stakeholder participation as limited to the participation of the public or affected community in general. It defines two types of public participation in the EIA process, participation of the public or community in the EIA process (informal participation) and participation of the public representatives in the EIS review process (formal participation). In this decree, the public is given rights to access information in the EIA process, to provide suggestions or comments on the proposed project, and to be members of the EIS review committee. Under the decree, the sector of government should inform the public on activities of the proposed plan and the EIS, compile and process all public inputs and present them to the EIS review committee, and facilitate public participation in the EIA process. The initiator of the project should inform the public on development activities before EIS is composed, consult the public affected in the EIA study, and inform the public on the EIS and the proposed measures for impact mitigation and impact monitoring. Stakeholder participation based on decree no. 08/2000 is mostly considered as informal participation in which participation is not obligatory, in which the guideline is mostly giving a suggestion to the initiator and to the sector of government to involve the public. The only formal participation is in reviewing EIS in which the stakeholders are given the opportunity to review EIS through their representatives. The new decree no. 08/2000 on stakeholder participation gives stakeholder participation a more active role compared to previous law.

In this research, respondents of the technical teams were asked to comment on stages of the EIA process and stakeholder participation for each stage as regulated by law and what happens in practice. The stages of the EIA process, as seen in table 6.1, are screening, scoping, analysis and prediction, reporting, review, decision-making and follow up. The respondents' general view is that all stages of the EIA process were being implemented in general practice, except the follow-up stage. Two of four respondents reported that the screening stage, which involves the determination of whether or not a project proposal requires an EIA study, had not being implemented. A possible reason for this is that screening was not a formal requirement, as some of the activities required by the EIA regulations were carried out within offices of the government sector. The respondents reported the follow up stage that consists of impact mitigation and monitoring as well as auditing or evaluation of the EIA as moderately implemented in practice. This suggests that the respondents view this stage as moderately important. The respondents were commonly only involved in the EIS review process, and never in the follow up stage. Although legislation has included this stage as part of the EIA process, the respondents did not seem to know how to implement since it was not within their responsibility. Moreover, the legislation has not mentioned an institution responsible for the monitoring process in the follow up stage (section 5.2). The respondents were commonly only involved in the EIS review process, and never in the follow up stage. Sadler and Verheem (1996) who reported a similar condition in most of the countries consulted, point out as a main weakness that although monitoring was a requirement in slightly over half of the jurisdictions consulted; impact management, monitoring and audits were less widely undertaken.

With respect to the role of key parties, as seen in tables 6.1 and 6.2, most respondents thought that the competent authorities from the government sector should be involved in all stages of the EIA process. Most respondents reported that a committee of experts should be involved in reviewing and in decision-making, in line with what was required according to the regulations. The Industry respondents reported that the initiator, as the owner of the project, should be involved in all stages of the EIA. The Sea Communications respondents contradicted this view, contending that the initiator should be involved only in the reporting stage. While the environmental authorities, the Ministry of Environment and the EIA office (*Bapedal*) should be involved in most of the EIA process. Similarly, the environmental experts should also had more participation. Sadler and Verheem (1996) recorded similar trend based on their study on the country status of the EIA. They reported that a majority of countries consulted had provisions regarding the role and responsibilities of environmental authority. This institution, in general, played a major role at the screening / scooping and review process stages. However, this role extended throughout the EIA stages only in a few countries consulted.

Regarding the public participation, majority of respondents reported that regulation was in place for stakeholder participation in the EIA, this indicated that all respondents were aware that there is a new law on stakeholder participation. The Industry respondents reported that public or local stakeholders should involve in most of EIA stages while the Sea Communications respondents were of the view that public involvement in all stages was unnecessary. In this respect, the respondents from the Industry sector had a more awareness of the law on stakeholder participation in the EIA process compared to those of the Sea Communications sector. The respondents understood the concept of stakeholder participation in EIA process and in the decrees as passive participation and not as active participation in all of the EIA stages. A possible reason for this might be that the new law on stakeholder participation was not implemented in the EIA process when this research was carried out and the technical teams were not yet familiar with the new law during this period. On the question on the type of participation of stakeholder, most respondents reported that participation should be in the form of providing information during field survey and consultation in the EIS compilation. This suggests that awareness of the technical team respondents on active stakeholder participation was limited. The technical teams of the institution sector at national level commonly followed the traditional decision-making process in this sector, which was controlled by the national government with limited stakeholder participation. Respondents' perception on EIA stages and stakeholders involved is shown in table 6.1, while respondents' perceptions on final responsible key parties for activities of the EIA process are presented in table 6.2.

Table 6.1. Technical team perceptions on EIA stages conducted and the stakeholders involved (n=4).

		Conducted		**Groups involved in the practice**				
Stages in the EIA process		**Yes**	**No**	**CA**	**EA**	**PU**	**EC**	**IN**
1. Screening		2		2				
2. Scoping		4		3	4	1	1	2
3. Analysis/Prediction		4		2		2	3	1
4. Reporting		4		2	2		1	2
5. Review		3		2	3	2	3	2
6. Decision-making		3		2	3	2	3	3
7. Follow up:	a. Mitigation measures	2	1					2
	b. Monitoring	2	1	2			1	2
	c. Auditing/Evaluation	2	1	2				1

Note: CA = Competent authority, EA = Environmental authority, PU = Public, EC = Environmental experts, and IN = Initiator.

Consistent with the previous answer, most respondents believed that the competent authority should be responsible for all the stages of the EIA process. The respondents viewed the initiator and competent authority as the final responsible parties in most of the early EIA stages. The review of EIS, a later stage, was reported by the respondents to be the responsibility of key parties other than the initiator. The approval of EIS and the issuance of permit should be the responsibility of the government sector. It can be concluded that the competent authority is viewed as being responsible for all activities while the initiator or the developer is viewed as only responsible for composing the EIS. The environmental authority, the office of the Ministry of Environment, the EIA office (*Bapedal*) and the expert committee should be involved only in EIS review process, according to the views of the respondents. In contrast, the Indonesian Ministry of Environment had given a legal status to almost all the EIS stages except the prediction and reporting stages, as reported by Sadler and Verheem (1996) based on their study on the country status of the EIA. The government sector, as the competent authority, had a more important function compared to the EIA office based on the view of staff of government sector as members of the technical team.

Concerning the sanction, most respondents reported that although sanctions for the violation of EIA decisions were authorised by ministerial decrees and government regulations, in practice however, implementation of law enforcement was limited. The respondents specifically mentioned that the type of sanction should be clearly stated in the regulations so that they could be used for law enforcement by district government.

Table 6.2. shows final responsible key parties for activities of the EIA process.

Table 6.2. Technical team perceptions on final responsible key parties for activities of the EIA process (n = 4).

Activities of the EIA process	Initiator	Competent authority	Environmental authority	Experts	Other
1. Use of the EIA procedure	3	3	1		related department
2. Propose the EIA	3	4	1		
3. EIA scoping	1	2	2		
4. EIS compilation	3	1			
5. Review/evaluate the EIS	1	4	3	3	
6. EIA approval		4			
7. Final decision-making		3	1	1	

6.3. Stakeholder Participation in the Coastal EIS Review Meetings

The assessment of coastal EIS review meetings for this research is based on law no. PP 51/1993 that gives the authority for EIS approval to the national level of sectors of government. The recent law no. 27/1999 that gives the authority of the EIA review to the local level was yet to be implemented when this research was carried out (1999-2000). Based on law no. PP 51/1993, the review of EIS was carried out by the central commission with the help of the technical team.

The membership of the technical team comprises professionals and experts related to the EIA and not institutions. The membership of the central commission represents decision-makers of institutions and consists of:

- permanent members (central and provincial levels), and
- temporary members.

The permanent members for the national level are the representatives of concerned sectors, the Board of Coordination for Investment (*BKPM*), the EIA office (*Bapedal*), the Ministry of Internal Affairs (*Depdagri*), and the National Board of Land

Registration (*BPN*). The permanent members for the provincial level are the representatives of central members at the provincial level, the District Planning office (*Bappeda*) and the NGOs. The temporary members are decided by the chairman of the central commission based on possible impacts generated by the proposed development plan.

For this research, the assessment of stakeholder participation in the coastal EIS review meetings is divided into:

- participation of stakeholders at the meetings, and
- input from stakeholders to the meetings.

Furthermore, the process of the coastal EIS review meetings will be assessed based on assurance of its objectivity as summarised by Wood (1995) and Mostert (1995). This includes the accreditation of EIA review committee, the setting up of an independent review body, the publication of the results of the review, the participation of the public, and the use of the review criteria in the review process, free access and active dissemination of information, the use of public meetings, public hearings and public tools of participation.

6.3.1. Participation of stakeholders during the meetings

Stakeholders who should participate in the coastal EIS review meetings according to law no. PP 51/1993 are:

- central commission, as the review committee,
- the technical teams of the concerned sectors,
- representatives of other government offices,
- the community, NGOs and universities,
- experts, and
- the district governments concerned.

Based on observation, participation of representatives of government institutions as members of the central commission (7-10 people) and the technical team (10-15 people) were sufficient for all meetings. The participation of 2 experts in the Sea Communications meetings was regarded as sufficient. The participation of consultants in all observed meetings was considered as highly sufficient (approximately 7-10 people per meeting). It is understood that they present the EIS on behalf of the initiator of the plan. The district governments did not participate in any of the meetings observed, although invitations were sent to them beforehand. However, the district government sent a letter to the Sea Communications meetings objecting to changes made in local policy on the prohibition of sand mining activities. The chairman of the meetings overruled the objection from the district government because the objection was unrelated to the purpose of the meeting, which was to review the content of the EIS. A participant from the EIA office said that the objection from the district government would be referred to a higher decision-maker. Participation of the community, universities and NGOs, as independent parties, was low or nil, with no participation from the community and NGOs in all the observed cases, and only one university attending one of the Industry meetings.

Table 6.3 presents a summary of participation of stakeholders in the 4 coastal EISs review meetings, as observed.

Table 6.3. Participation of stakeholders in the 4 coastal EIS review meetings.

Review items	**The Industry sector**		**The Sea Communications sector**	
	Meeting 1	**Meeting 2**	**Meeting 3**	**Meeting 4**
1. Number of attendees				
a. District governments				
b. Related institutions	●	●	●	●
c. Technical team	●	●	●	●
d. Community/Private/NGO/University				
e. Experts			○	○
2. Giving comments				
a. District governments				
b. Related institutions	■	■	■	■
c. Technical team	■	■	■	■
d. Community/Private/NGO/University				
e. Experts			■	■
3. Participation of consultants				
a. In giving presentation	□	■	■	■
b. In giving answers	■	□	■	■

Note: ● Enough participants, ○ Limited participants
■ Enough time, □ Limited time

Concerning the participation of stakeholders in the discussions, all participants were given an opportunity to make comments or ask questions at all the meetings observed. Priority was given here to participants outside the concerned sectors like other government institutions. Technical teams were then given the opportunity to review the documents. About 10 members of the technical team took part in the discussions and were given enough time to express their views and concerns and ask questions. Thus, they played a dominant role in reviewing the reports at all the meetings. In the Sea Communications sector meetings, the consultants were given enough time to make presentations. In the Industry meetings, the secretary of the technical team made presentations to the central commission on behalf of the consultants, a practice within the Industry sector to save time and expedite work. Comments from members of the central commission who were not members of the technical team were not given high consideration. The participants from other government institutions, although having enough knowledge to comment or review the technical aspects of the reports, confined their comments to the role of their institutions.

6.3.2. Inputs from stakeholders to the meetings

In all the observed meetings, there was sufficient input from the consultants, the technical teams and experts of the Sea Communications sector. Inputs from the technical teams in all cases and from the experts at the Sea Communications meetings are given special consideration in reaching the final agreement. But inputs from related institutions were not given much weight in the decision-making at the Sea Communications meetings. Most of the participants outside the technical team who were representing their offices were not considered as decision-makers. No input was received from the district government and independent parties (community, NGOs, or universities) at all the meetings, as they were not present at the meetings. The decision-making authority at the meetings was with the chairman. Concerning agreement at meetings, consensus between consultants and initiators was low, especially for the Industry meetings. At all meetings sufficient agreement was achieved between technical teams and other related institutions. In the final stages, the experts and the technical teams at the Sea Communications meetings reached sufficient agreement. However, the entire coastal EISs were not directly approved at the meetings, as it was decided that either the report should be revised independently by the consultants, as in the Sea Communications meetings, or revised in consultation with the technical team, as observed in the Industry meetings.

There are differences in the methods of EIS approval of the two sector offices:

- In the Sea Communications sector, the chairman of the meetings sent the recommendations on the EIS to the Director General for Sea Communications and then to the Minister for Communications for the approval of the EIS.
- In the Industry sector, after sufficient revision and improvement by the consultants, the chairman of the central EIA commission for the Industry meeting will approve the reports on behalf of the Secretary General of the Minister for the Industry.
- The responsibility of EIS approval for the Sea Communications sector was with the Director General or the Minister for Communications, while for the Industry sector, it was with the chairman of the EIA central commission.

Table 6.4 presents inputs from stakeholders to the 4 coastal EIS review meetings.

Table 6.4. Inputs from stakeholders to the 4 coastal EIS review meetings.

Review items	The Industry sector		The Sea Communications sector	
	Meeting 1	Meeting 2	Meeting 3	Meeting 4
1. Inputs from:				
a. Consultants	■	■	■	■
b. Technical team	■	■	■	■
c. Related institutions	■	■	■	■
d. District governments				
e. Experts			■	■
f. Community/Private/NGO/University				
2. Perception of agreement				
a. Consultant and initiator	□	□	■	■
b. Technical team and other institutions	■	■	■	■
c. Experts and technical team	□	□	■	■
3. Group inputs for decision-making				
a. Technical team	■	■	■	■
b. Related institutions	■	■	■	■
c. District governments				
d. Experts			■	■
e. Community/Private/NGO/University				

Note: ■ Enough input □ Limited input

6.3.3. Objectivity of the coastal EIS review meetings

Based on the methods used to assess the objectivity of the review process as promoted independently by Woods and Mostert in 1995 (section 6.3), the analysis of the four coastal EIS review meetings observed are as follows.

- The accreditation of EIA review committee of the Indonesian EIA system was not yet known. Therefore the membership of the EIA review committee that consisted of permanent and temporary members was not finalised. However the permanent members at national level had been appointed. They had been drawn from representatives of concerned sectors, the EIA office, Ministry of Internal Affairs, and the National Board of Land Registration. The permanent members from the provincial and local level as well as the temporary members will vary according to the location and type of projects and possible significant impacts generated by the projects under assessment. However, there is a requirement for members of the technical teams and consultants to work on the review process and for the consultant to conduct the environmental impact assessment and to compile the EIS. The requirements is in the form of issuing a certificate for basic EIA training and an advance EIA training for practitioners who will prepare the EIS in addition to an advance EIA training for evaluator which is intended for members of the EIA review commission.
- The setting up of an independent review body was not recognised in the Indonesian EIA system. The review of the EIS, based on law no. PP 51/1993, was carried out by the central commission with the help of the technical team.
- The use of standard review criteria in the Indonesian EIS review process was not available at the time this research was carried out (1999-2000). The available

method to review the EIS during that period was in the form of a review checklist that needed to be filled-in with comments of the review members based on their sector of activity. A review mechanism was set up in year 2001 in the form of guidance on the contents of EIS. It was not in the form of standard criteria to review quality of EIS.

- Free access and active dissemination of information prior to conducting the four EIS review process under consideration was not sufficient. For example, the EISs of the Sea Communication meetings were distributed one week prior to the meetings, and only three days prior to the Industry review meetings. The limited time for EISs distribution hampered the stakeholders in particular from district level to access the intended information. Thus, the dissemination of the EIS and other supporting information was limited only to the participants of the review meetings.
- The publication of the results of the review in Indonesian EIA system was limited as reported by Sadler and Verheem (1996). The result of the review process was not published in the form of advisory document intended only for internal use. Later a bound copy was published as limited edition.
- It was observed that the participation of the public at the four meetings was inadequate. Neither the public nor the local government attended as the local stakeholders. The letter of objection of the district government for the Sea Communication meetings was not taken in the decision-making, and it was only 'taken into consideration'.
- Tools of public participation in the form of public meetings, public hearings and public participation in addition to the four EIS review meetings were not conducted. It is not known whether these were made use of prior to the EIS review.

High objectivity and thorough reviews of EIS will have a direct and positive effect on the quality of the EIS reviewed and thus the quality of the decision-making. Based on the assessment of the four coastal EIS review meetings using the methods promoted independently by Woods and Mostert in 1995, it is proposed that the objectivity of the four review meetings need a lot of improvements. In general, most of the methods used to review the objectivity of the EIS review meetings shows inadequacy. There were several constraints commonly appearing in the EIS review process as reported by Scholten (1995). Statutory constraints may exist, for example, in the absence of a formal requirement in the legislation with proper time limits and indications for participating stakeholders, or constraints experienced in the case with impacts involving different jurisdictions. Methodological constraints exist in the non-availability of formal checklist and the EIS review criteria. The capacity problems occur if there is a lacking of sufficiently trained EIA managers and experts in the EIS review process.

6.4. Stakeholder Participation in the Pre-Coastal EIS Compilation

The stakeholder participation in the pre-coastal EIS compilation is assessed based on decree no. 08/2000, in which the public is given rights to access information in the EIA process, to provide suggestions or comments on the proposed project, and to be members of the EIS review committee. In this research, the assessment of stakeholder participation in the pre-coastal EIS compilation is illustrated by a case study of the development of the Jababeka Industrial Estate that involving a reclamation activity. The assessment is based on perceptions of stakeholder in regard to:

1. Their participation in project approval, communication with the developer of the project, and information input to the project, and
2. Their access to information given in the project, written reports and other information expected from the project.

6.4.1. Participation of stakeholders in project approval

Participation of stakeholders in project approval can be measured from their familiarity with the plan, their involvement in project approval, and their opinions of the plan. In the case of the Jababeka Industrial Estate Project, overall familiarity of respondents to the project was judged as moderate. Less than half (46%) of all respondents reported that they either knew the project very well or knew something about it while more than half (55%) reported that they did not know about the project at all. This level of familiarity of respondents with the project also applied to groups of participants, the district government and the community respondents. Almost half of the groups felt they knew the project and the other half reported that they did not. Out of those claiming knowledge of the project, the district government respondents were possibly from the District Planning Office, while the community respondents who knew about the project were possibly those whose lands were sold to the project. It was observed that in this case study no information concerning the project was given directly to the respondents. The community respondents reported that they received undocumented information on the project information from their community groups. This led to half of total respondents feeling unfamiliar with the project as seen on figure 6.1. This figure would suggest that the familiarity of the respondents to the projects was moderate, and the spread of information about the project could be considered as low to moderate. This source of information showed the moderate level of familiarity of respondents on the project. It provides a significant evidence of the trends of opinion among the respondents towards the project. Majority respondents (83%) reported that they either opposed the project or did not have opinion concerning the project, in contrast to a small percentage of respondents that supported the project. This figure also applied to the majority of each group of respondent that opposed the project as seen on figure 6.2. Those who believed that they supported the project were mostly the district government respondents, while those who believed they opposed the project were mostly the community respondents who felt they were not being informed on the project. This would suggest that the low level of dissemination of information on the project lead to the low level of perception on supporting the project. Figure 6.1 presents the overall familiarity of respondents with the Jababeka project and figure 6.2 presents the opinions of respondents toward the Jababeka project.

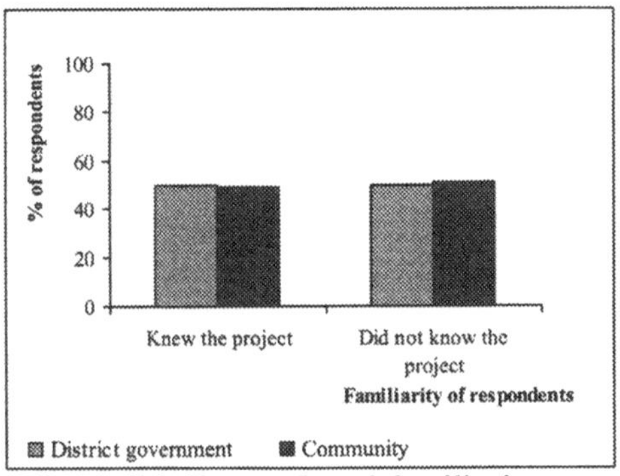

Figure 6.1. Respondents' familiarity with the Jababeka project (n = 86).

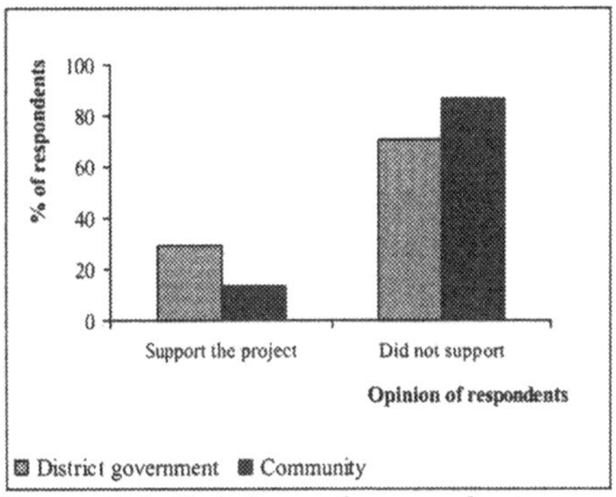

Figure 6.2. Opinions of respondents toward the Jababeka project (n = 86).

As on figure 6.3, most of the respondents (92%) did not feel they were involved in project approval. Less than half (40%) of the total district government respondents reported involvement in the approval process, and possibly this number came from the respondents whose offices related to the approval process of the project such as the District Planning Office. But the level of participation of the community respondents was very low with almost all (92%) reporting that they were not involved in any aspect of project approval. These figures suggest that the level of stakeholder participation in project approval was very low. These figures of participation changed somewhat when respondents were asked about the "desired" level of involvement or about how they believed they should be involved in development approval. As seen on figure 6.4, approximately 74% of all respondents consisting of community (74%) and district government (89%) respondents reported that they should be involved in the approval process. The community respondents reported that they would provide some input and information if they were involved in the project approval. The high level of desire to participate among the respondents indicates their awareness of the importance of participation and of giving information or input. However, in practice they were ignored in the project approval. This suggests that the empowerment of stakeholder on participation in the project approval was lacking. Figure 6.3 presents the perceptions of respondents about their participation in project approval, while figure 6.4 presents the perceptions of their desire to participate in project approval.

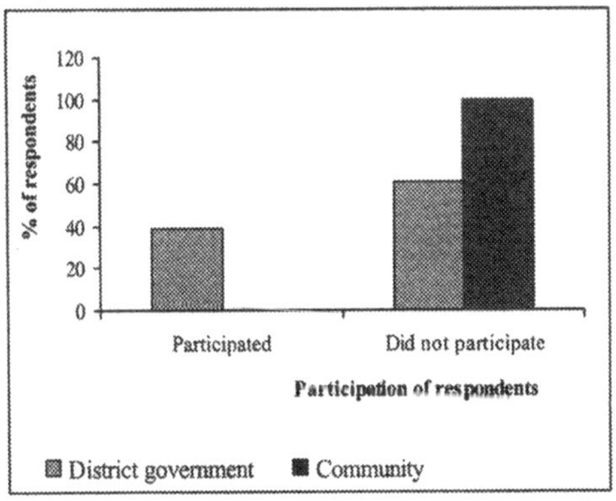

Figure 6.3. Respondents' perceptions of their participation in Jababeka project approval (n = 86).

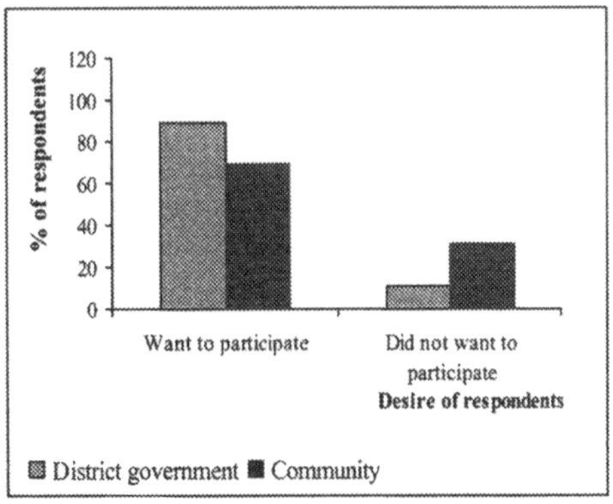

Figure 6.4. Respondent' perceptions of their desire to participate in project approval (n = 86).

Since only a small percentage (8%) of total respondents felt that they were involved in the approval process, therefore, only one fifth (17%) of all respondents expressed support for the project. As seen on figure 6.2, most of the respondents from the district government (71%) and community (86%) expressed their objections to the existence of the project. The objections ranged from no support to the project (45%) to not commenting on the project (39%). These figures suggest that most respondents did not support the plan. Some reasons for supporting or rejecting the project had been identified. Those who supported the project believed that the project would generate jobs (61%) and bring prosperity (11%), and would improve the environment (5%). While those who did not support the project believed that it would pollute the environment (51%) and decrease the productive land area (37%). The high level of objection by the community to the project had resulted in the presence of conflicts mostly due to problems on land acquisition (Tiwi, 1999). This suggests that the low level of stakeholder participation on the project approval had led to the low level of perception of respondents on supporting the project.

6.4.2. Perceptions of stakeholders on communication and information input

Respondents were asked whether they had any communication with other parties regarding the plan and for what reason. As seen on figure 6.5, about 71% of respondents from district government (28%) and community (72%) reported that other parties had contacted them regarding the plan. Staff of other government offices such as those from the local to the higher levels mostly contacted the district government respondents. In contrast, the contacts of the community respondents were from the village government staff and other parties such as informal representative of the project proponents for land acquisition. Although these figures suggest that in general, respondents had sufficient contacts regarding the project, most of the communication was conducted between the respondents with other parties other than the project proponent and in one direction from the communicator to the respondents. These figures suggest that in general, respondents had sufficient contacts regarding the project. However, most of the communication was conducted between the respondents with other parties other than the project proponent and in one direction from the communicator to the respondents. This was indicated by mostly officials of village administration (27%), community groups (29%) and district government (8%) that contacted the respondents while representatives of NGOs, research offices, national government, and mass media contacted small percentages of other respondents. These figures suggest that contacts were made with the respondents only at the lowest level of district government such as the village administration. This is understandable, as it is the level of district government closest to the community. Various methods had been used in communicating information about the project. These include self-organised methods (21%), methods organised by others (38%) and a combination of both (42%). These figures suggest that initiatives for participation in the project generally came from sources other than the respondents and mostly from top to bottom levels for the needs of other than the respondents.

Stakeholders were questioned on their reasons for their contacts with the project, whether they provided any input to the planning of the project. Majority of community and district government respondents indicated that giving suggestions and input was their major reason for their contact with the project in addition to giving information and data. For others the community respondents added that land buying was a reason for their contact with the project proponent. These figures suggest that there was an awareness of the respondents on giving sufficient input from the respondents to the project. However, there is a question regarding the quality of information and how the project proponent had taken much information into consideration.

A questionnaire on the respondents' communication with the project and on the tasks or roles of the project related to the environment gave a similar result. Most respondents (51%) from district governments (52%) and community groups (56%) expressed views and concerns related to environmental issues and jobs provision in their area as well as by giving information or suggestions to the project. These figures suggest that there was an awareness of the respondents on possible environmental impact generated by the project since it will utilise coastal and marine area as part of the community's fishing ground. The opinions of respondents about the ability of stakeholders to increase environmental awareness in any future cases of environment pollution were also obtained. The district government was the most popular choice of the respondents. They gave lower ratings to community groups (26%) and the national government (17%). The respondents also expected the village administration, NGOs, universities and research institutions, and mass media to participate in raising environmental awareness. This would suggest that the respondents were aware of the stakeholders that should be responsible for raising environmental awareness.

Figure 6.5 presents information on stakeholders who made contact with the respondents regarding the project.

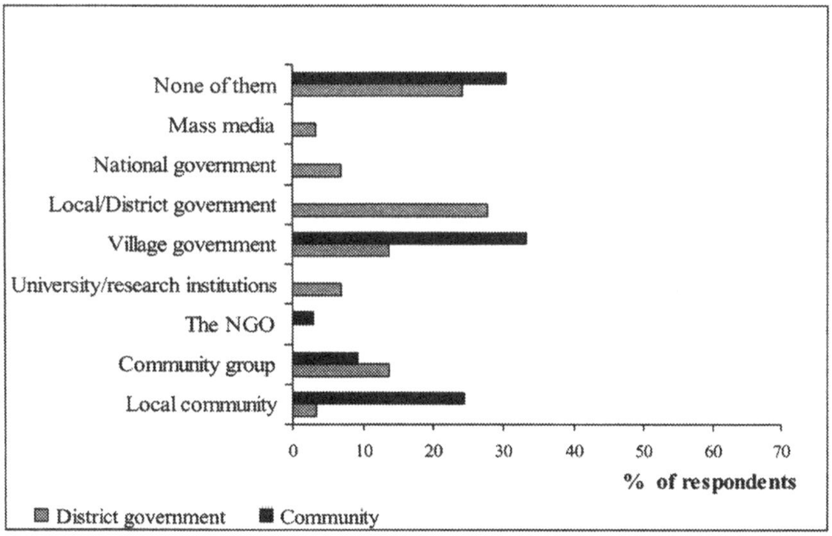

Figure 6.5. Stakeholders who made contacts with the district respondents regarding the Jababeka project (n = 86).

6.4.3. Perception of stakeholders on access to information

Respondents' perception of access to information sources concerning the project involves awareness of information given by the project, information in written reports, and other expected information from the project.

Information given by the project

Information about the project was provided by direct communication of information given by the project, through announcements, and through visits to project sites.

As seen on figure 6.6, the respondents indicated that direct information sources (58%), and announcements (48%) provided the most useful and important information. Visits to project sites and announcement from the project were considered unimportant in providing information by most of the respondents, especially respondents of the community. Compared to other types of information listed in the questionnaire, the respondents gave higher preference to direct information given by the project instead of written information. This suggests that majority of the respondents expected open communication with the project proponent, preferably by direct information. In practice, the case study project had never released any information directly to the respondents and no visits to sites had been made. In the absence of direct communication of information, most respondents expected to have access to information directly from the project.

Figures 6.6 and 6.7 present the perceptions of Serang district government and community respondents on the importance of direct information sources.

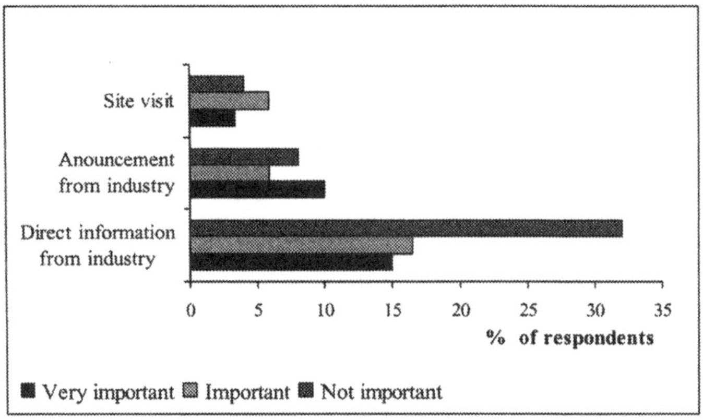

Figure 6.6. Perceptions of Serang district government on the importance of direct information sources (n = 18).

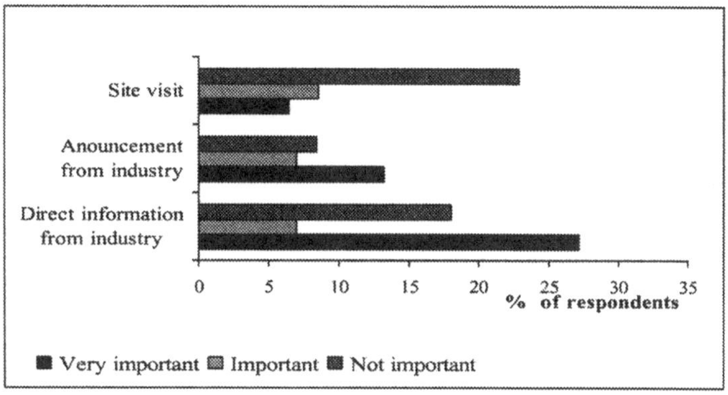

Figure 6.7. Perceptions of community respondents on the importance of direct information sources (n = 68).

Information from respondents on written reports

The respondents believed that the most important sources of information regarding the case study were sources at various levels of district government: the district, sub-district and village government offices. This was because these offices had more access to the development plan compared to other sources of information. Respondents were also asked about their perceptions of access to information through written reports: development plan reports, the EIA reports, and the NGO reports. Of these, the government report was believed to be the most important source of information and the development plan the second important. The respondents believed that the other reports were of moderate importance as sources of information. This would suggest that the government especially at district level was considered by the community as the highest important source for documented information. The level of importance ascribed by the stakeholders to the written reports can be used to measure the level of confidence in the reports. The district government (56%) ranked the development plan as the most important source of information, as they had more access to the report on the development plan compared to the community. The development plan and the government report were given preference as the most important sources on written information, while EIS was considered as a moderately important source of information as seen on figure 6.8. A possible reason for this might be that EIS was still considered a confidential report and the distribution of information within EIS was limited. This suggests that awareness on the importance of EIS needs to be improved. The respondents did not consider NGO reports and newspapers as important sources of information, and ranked them as the lowest important information sources.

Figure 6.8 presents the perceptions of respondents on the importance of written reports as sources of information.

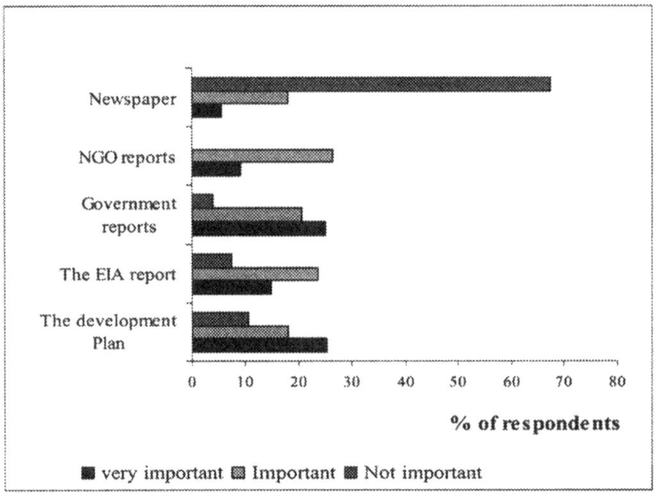

Figure 6.8. Perceptions of respondents on the importance of written reports (n =86).

Other information expected from the project

The respondents expected other information outside of the questionnaire. The respondents, who were dissatisfied with the information they received, desired information including: labour requirements, especially the required numbers and the localities from which recruitment will be made (22%); time of commencement of construction of industry (25%); and the waste treatment plan (17%). Information on construction time and labour was highly desired because development of industry will create many new jobs and also because most of the labour in other industries were recruited from outside areas. Often the labour skills available locally did not match the requirements of the industries. The expected information on type of industry within the industrial estate and the waste treatment plan was related to pollution. It was important to the respondents because, several years ago, they experienced river pollution from industries in the study area that devastated activities of the fishery sector. This suggests respondent's lacked of awareness on stakeholder participation in the guidelines as well as in practice.

6.5. Conclusions

Research question 7. How is the participation of stakeholder in the EIA process based on perceptions of technical teams?

The technical teams, as respondents, were questioned about their perceptions of stakeholder participation based on regulations and what happens in practice. The respondents reported that all stages of the EIA process, except the follow-up stage, were being implemented in practice. Majority respondents reported the follow-up stage that consists of mitigation, monitoring, and auditing or evaluation, as moderately implemented in practice. On the question, Who has the final responsible for implementing the stages of the EIA process? majority respondents gave opinions, as: the competent authority should be responsible for all activities while the initiator was only responsible for composing the EIS. The competent authority was viewed as a dominant force while it was suggested that the environmental authority and expert committee should be involved only in the EIA review process. The approval of the EIS and the issuance of permits should be the responsibility of the sector of government as indicated in the law enforced at that time.

The new decree no 08/2000 on stakeholder participation gives stakeholder participation a more active role compared to previous laws. However, the decree was not clearly address the guidance on stakeholder participation, and was only advocate the initiators or the developers to only inform the public concerning the EIA process. The respondents though aware of the new decree on stakeholder participation, were, however, unfamiliar with its concept. Therefore, some respondents reported that the public or local stakeholder should participate in all stages of the EIA as had been stated in the new decree. According to other respondents, the local stakeholder should participate only in providing input and information during the field survey or only during the EIS-review meeting. In general, all respondents viewed participation of stakeholders in all stages of the EIA process as passive. This would suggest that the respondents were lack of awareness on stakeholder participation both in the guidelines and in the practice. The technical team was not familiar with this decree when the research was being conducted (1999–2000) and it reported that this decree was not public knowledge during that period. Hughes (1998), based on his studies on EIA guidelines and stakeholder involvement, reported that there were a number of common features that appear to undermine their value as guidance tools for stakeholder participation. Among these were lack of clarity, confusing use of terminology, lack of practical guidance, and lack of proactive support for stakeholder participation in the EIA process.

Research question 8. How is the participation of stakeholder in the coastal EIS review meetings?

Based on observation, the participation of representatives of government institutions, the technical teams, and consultants at all meetings were sufficient, while the participation of local stakeholders was low. The district government, community, universities as well as NGOs, as independent parties, were absent at all the meetings observed. Participation of stakeholders in the discussions was considered as sufficient, as observed in all the meetings. Comments on the reports from the technical team at the meetings were considered as sufficient, while comments from other government institutions, as members of central commission, were considered insufficient. The comments from members of the central commission outside the technical team were considered insufficient, since most of their comments were limited to the role of their institution although they are competent to comment on the EIA. The consultants and the technical teams provided enough input to all the meetings, as did the experts in the Sea Communications meetings. Input from participants outside the technical team was considered limited since most of them were not decision-makers. The perception of agreement among participants at the meetings was that: agreement between consultants and initiators was "low," especially at the Industry meetings, and agreement between the technical teams and other related institutions was "enough" at all meetings. Final agreement on the EIS was not for the approval of all the aspects of the EIS, as recommendations were made that some parts should be revised independently by the consultant or under the supervision of the technical teams. The responsibility of EIS approval for Sea Communications was with the minister while for Industry it is with the chairman of the meetings.

Based on the assessment of the four EIS review meetings using the methods promoted independently by Woods and Mostert in 1995, it suggests that the objectivity of the four review meetings need a lot of improvements. In general, most of the methods used to review the objectivity of the EIS review meetings show inadequacy. The accreditation of EIA review committee in Indonesian EIA system had not been made known yet. In addition, the Indonesian EIA system did not provide for setting up of an independent review body to review EIS. It was based on law no. PP 51/1993 and was carried out by the central commission with the help of the technical team. The shortcomings of this coastal EIA review meetings were:

- Free access and active dissemination of information prior to the conduct of the four EIS review process was not sufficient.
- The publication of the results of the review in Indonesian EIA system was limited. The result of the review process in the form of advisory document was not published, while the results in the form of formal binding document were published later in limited copies.
- Poor participation of the public in the four meetings was inadequate as observed. The utilisation of public meetings, public hearings and public participation tools in addition to the four EIS review meetings was not conducted.

Promoting objectivity of EIS review is one of the most difficult areas in this process. There are various methods promoted to ensure objectivity of EIS review as suggested by Wood and Mostert in 1995 (section 6.3). Several strategies also have been suggested by Scholten (1995) to improve the effectiveness of EIS review process. These include, setting the boundaries of the review process, selecting the team to review, utilising input from stakeholder participation, using the review criteria, carrying out the review process through stages, publishing the review report, and determining remedial options if serious deficiency occurs.

Research question 9. How is the participation of stakeholder in the pre-coastal EIS compilation?

In evaluating the quality of stakeholder participation in the pre-coastal EIS compilation, respondents were questioned on their:

- familiarity with the project,
- perceptions of participation in the project,
- inputs to the project,
- access to information about the project, and
- perceptions of written reports.

As illustrated in the Jababeka Industrial Estate Project, in general, respondents reported that they knew about the project. This suggests that accessibility of information about the project was moderate. Thus the respondents' overall familiarity with the project was judged as moderate. Concerning participation, most respondents felt that they were not involved in project approval. Almost majority of respondents did not support the project, with responses ranging from not supporting the project to not having an opinion about the project. Thus the level of stakeholder participation in project approval was very low. In regard to communication of project information to the respondents, most respondents reported that other parties like village administration and community groups contacted them. However, respondents reported that their access to information about the project was limited as they mostly received news and announcements given indirectly by the project. Written reports were considered as the second most important source of information, and respondents rated the government report as the most important one, the development plan as the second important report, while the EIA report was considered as moderately important. This suggests that the level of confidence of the stakeholders to the EIS be on average and need to be improved.

Based on the analysis of stakeholder participation in the previous sections and stages of participation suggested by Soet (1987) as well as 4 types of participation distinguished by Bisset (2000) as discussed in section 2.3, it could be concluded for this research that:

- The stage of stakeholder participation in the pre-coastal EIS compilation as illustrated in the Jababeka Project approval was "limited and controlled participation". The participation of the public was very limited, in which the utilisation of public meetings, public hearings and public participation tools was not exist. Decisions were taken unilaterally by the national agency concerned with limited participation of the district government even the community.
- The degree of participation was only limited to "information dissemination", in which the degree of participation was more towards passive participation, since the flow of information was more a one-way process from the project proponent to the public and was not given directly by the project. Free access of information by the respondents was not achieved, in which the publication of the information was considered limited. The respondents received information from the source other than the project proponent or the district government. The existence of conflicts between the community and the project on land acquisition among others indicated that information being used for decision-making were poor
- The empowerment of stakeholders on participation was lacking. Awareness of the respondents on possible environmental impacts generated by the projects as well as awareness of giving their input or information to the project was sufficient, though, it was being ignored in the pre-EIS compilation. The dissemination of information on the project was limited, and the participation of public on project approval was lacking.

7. DISCUSSION

7.1. Introduction

On the basis of research results as presented in chapters 5 and 6, this chapter presents discussions on the use of environmental information and stakeholder participation in the coastal EIA process. Discussions are presented as on the following main topics:

- availability of priority information,
- information sharing,
- information management,
- participation of stakeholders,
- empowerment of stakeholder participation, and
- future direction on stakeholder participation.

7.2. Limited Availability of Priority Information

In order that proper decision making on coastal EIA and also cost-effective policies and development plans are developed accurate information and rational analysis are required. Thus, information plays a very critical role, in helping to build a consensus toward ensuring compliance between development plans with environmental policies and regulations. In this research, most of main information sources for the coastal EIS such as environmental data, status of natural resources, and environmental monitoring data were used by a majority of the consultants, technical team, and district government respondents. It can be considered as a good starting point for the coastal EIA. However, most respondents were not satisfied with the available information sources. As seen in tables 5.3 and 5.4, only one of the five most needed information sources was found to be one of the most available information sources, and three of the five most available sources were found to be within the five least relevant information sources. Information sources related to methodology and technical information were better utilised compared to sources related to local data, since their availability was considered as high to moderate. On the other hand, local data such as monitoring data and status of natural resources were considered scarce, although this information was highly needed for the coastal EIS. Environment data was rarely used to draft and to review EIS.

The consultants were not satisfied with the availability of important sources to draft EIS as they have only limited access to feasibility data. The relevant data were still undergoing revision or were considered confidential by the initiator and consultant who conduct the feasibility study. It was difficult to obtain environmental data either from the initiator or the district government. If he was not supported by necessary information and facilities to measure the data, especially marine data, to assess the key impacts generated by the proposed plan, he would use available data, or to obtain primary data accordingly, to draft the coastal EIS. A weak aspect of the EIA guidelines was the biological components of the marine area. Since the guidelines provided limited details on indicators of parameters within the biological component, the activities referred to in the guidelines such as drafting and reviewing of reports and monitoring lacked marine information. Concerning environmental information in the environmental monitoring activity, it was widely known that baseline environmental data were limited to the district government level. District government officials reported that they have only limited access to information at district level where necessary information was still handled by sector offices, and this information was rarely shared with other offices. Big projects at national level usually had sufficient baseline data, but the data were located in the central offices. Research studies that were commonly considered as partial research produced limited information on all aspects of the environment. Basic information necessary provided by the coastal EIS was unavailable and, even if available, often difficult to use as baseline data. The level of confidence to the coastal EIS was considered as an average. The unavailability of baseline data had resulted into limited monitoring activities. The research identified the following constraints to environmental monitoring activities; lack of resources in data collection, limited laboratory facilities, limited capacities of human resources, low capacity building related to data management, and unavailability of institutions authorised to conduct monitoring activities. These constraints on monitoring the environmental impacts were getting stronger. Furthermore, since law enforcement was lacking, the environmental monitoring activities seem unimportant. Similar problems were experienced in several other countries (Scholten, 1995) where environmental monitoring activities were deficient, resulted on the limited availability of environmental information. Similarly, Biswas (1992) reported that lack of follow-up monitoring was experienced in developing as well as developed countries. EIA as it was practised seems to end immediately after it was approved and after the environmental clearance of the project had been received.

Accurate data on environmental conditions and trends in Indonesia were not readily available as reported by World Bank (1993). The two main sources of published environmental information were the State Ministry of Environment (SMoE) which published data and trends on environmental conditions and trends annually and the Central Bureau of Statistics (*BPS*) which annually produced a report of environment-related statistics. However, the available data of these sources was based on administrative boundaries, and was not easily related to ecosystem boundaries. Furthermore, World Bank (1993) also noted that very little research had been carried out on interaction of human activities on the environment. Similarly, Asia Foundation (2002) added that the provision of local priority information was commonly hampered by the lack of co-operation and information sharing among the local governments in Indonesia. Furthermore, Dahuri and Dutton (2000) noted those lack of knowledge and awareness of coastal and marine resources as well as lack of institutional capacity on coastal and marine management in Indonesia hampered the provision of priority information.

7.3. Limited Information Sharing

Discussion of the availability of priority information indicated that information sharing is crucial. The research identified the cause of lack of information was the limited information transfer between offices at district level and between district offices and offices at provincial or national levels (section 5.5.2). Respondents reported that the flow of information or exchange of information between offices was considered to be unimportant for the coastal EIS. The importance of information sharing was considered to be limited. The respondents mostly used the limited flow of information within one's own office rather than prospectively bigger flows of information between offices. The flow and sharing of information between offices at local, central and provincial levels and even within the same office, which was important for the compilation and review of coastal EIS and coastal management, was limited.

The district government had limited access to local information on marine resources (section 5.5.2). Most of the local data, especially marine data, was found within the sector offices at national level, since most of marine projects were handled by the national government. Marine activities were co-ordinated and arranged by the Sea Communications office at the national level, and marine data that was categorised as "safety of sea transportation line" is confidential. The district government found it difficult to access the local data because there were no local regulations or a mechanism to facilitate access to the data distributed among sector offices. The procedures that need to be followed were time consuming. Access to information depends on to the willingness of the data holder to share the information. Some local data exist in research institutions and universities located in different cities. In the study area, research data were distributed among national and local research institutions, NGOs, as well as among the EIA consultants. Research institutions that hold data did not know to whom they should give the data, as there was no co-ordination of marine data management at the local level. Furthermore, the level of data held within the research institutions did not suit the needs of the district government.

Indonesian district governments commonly experienced the reluctance to share information, especially on coastal management. Dahuri and Dutton (2000) reported that the limited information sharing in Indonesian coastal and marine management among others was due to historically a traditionally sectoral project base, in which the sector offices commonly hold the data they collected which were generally considered as confidential. Furthermore, Asia Foundation (2002) had noted that lack of co-operation and sharing of information on some issues hindered administration and management of the Indonesian local government.

7.4. Poor Information Management

The district government, especially the District Planning Office (*Bappeda),* had the authority for information management at district level. But as these offices were the project bases for the national government, their activities related to information management were limited as user oriented. Local data collection was commonly related to big projects that need significant funding and which are generally conducted at national level. George (2000) noted that some of the environmental data gathered in developing countries might be hard to access because it was not public domain, and sources of available data may be poorly referenced. On the other hand, access to such sources depended on to the ability of knowing or finding who possesses the data.

Brown (1994), based on her research on the availability of information sources within local government in Australia, identified several shortcomings in the information management there. These include information overload, lack of information at the local scale, closeness of information system, use of modern information technology, fragmentation of responsibility, lack of appropriate research and investigation, lack of integrating skills, lack of access to local information, lack of resources, and lack of clear policies. These constraints will be used to discuss similar problems that possibly exist in the study area.

Information overloaded

As indicated by the research on information sources (section 5.4 and 5.5), most respondents who were consultants and district government staff indicated that training and courses were considered the most highly available source of information. In the case of information related to sectoral activities, some information was abundant in some sectoral offices it was scarce in others. The information that related to one sector of activity was exclusively owned by the office of that sector of activity as in the case of land registration and mapping in the study area (section 3.5.3).

Lack of information at the local scale

The research also noted that the availability of local information as environmental data, environmental monitoring data, and status of natural resources was limited (section. 5.4 and 5.5). The respondents reported that they were lacking of priority information on local scale whether to compose EIS or to review it. The environmental monitoring, as one provision of local information, was lacking of its activity. Most of parameters and indicators were seldom or never being monitored as reported by district government respondents (section 5.7).

Lack of access to local information and closeness of information system

One of the issues led to the limited availability of local information was the lacking of information sharing and the closeness of information channel among the stakeholders. Flows of information within and between offices were not important to compose and to review EIS as reported by respondents (section 5.4 and 5.5). They mostly used the information from their own. In particular, local information was commonly own by sector office at national level or research institutions/universities in the big cities. These data was commonly categorised as confidential and need some procedure for district government in order to access.

Fragmentation of responsibility and lack of integrating skills

As discussed in section 3.5, the conflicts in coastal and marine management were caused by fragmentation and distribution of responsibility for land and sea areas, permits and EIS approval among central and local institutions. Different consultants handling the feasibility study and the coastal EIA study had also contributed to the conflicts (section 5.4). The district government was traditionally influenced by sectoral development approach, while experience in regional autonomy was lacking. This contributes to the lack of integration in the resources of the district government.

Lack of clear policies and lack of resources
Often the institution that has the responsibility of collecting information lacked the capacity to do it as identified from the research (section 5.7). In the case of environmental monitoring, the major restriction was being no identified executing institution for monitoring activities. Most of constraints experienced by district government in this regard were related to funding, human resources, skills, data and analysis.

Lack of appropriate research
In Indonesia local data was difficult to obtain despite numerous efforts on research data collection conducted by international research programme. In the study area the only research programme conducted was the Teluk Banten Research Programme – a research collaboration programme between the Netherlands and Indonesia. The research focuses on the 4 important eco-systems.

Most of the constraints on information management as discussed occurred in the research area, therefore it is suggested that the information management within the district government was poor.

7.5. Limited Stakeholder Participation in the Coastal EIA Process

Traditionally, the Indonesian EIA system does not recognise stakeholder participation in the decision-making process. The concept of stakeholder participation was indirectly addressed by several regulations in force prior to decree no. 08/2000, in which the meaning of the stakeholder was limited to the public as the community in general. These regulations on stakeholder participation did not provide clear guidance on measures to enable participation and did not clearly specify the right of stakeholders to access information. Since the concept of stakeholder participation was traditionally not recognised, these regulations were more of an advisory and passive nature. The measures provided in the regulations were generally developer-led, in which the mechanism adopted tends to be biased towards the developer's interests. The notion of public participation, as defined in the decrees, was only translated into a situation where the public was being informed and not consulted. The level of participation was passive, which was directed in one way, from the proponent to the public, and participation did not influence the process of decision-making (section 6.3). Furthermore, limited guidance and measures to encourage stakeholder participation and the inaccessibility of the EIS documents to stakeholders limit their participation in the EIA process within the sectors of government dealing with the EIA.

In the practice, the research identified that the familiarity of stakeholders with the project, their participation in development approval and the communication process between the developer and the stakeholder concerning the plan were generally low (section 6.4). The stakeholder was only expected to participate in providing data and information during the field survey while they have limited access to information concerning the development plan and the coastal EIA process. The participation of stakeholders other than the competent authority in reviewing the coastal EIS was low and it did not influence the process of decision-making. The local stakeholders participate in the review process only after the reports had been finalised and it was unlikely that their views would be accommodated in the coastal EIS (section 6.3). Objectivity of the four coastal EIS review processes as observed were not satisfactory and need a lot of improvements (section 6.3). Based on the definition of stakeholder participation as suggested by Soet (1987: in section 2.3.3), this type of stakeholder participation as in the Jababeka project approval is specified as "limited and controlled participation." The available measures for accessing information was limited. The EIS reports and development reports as well as other information data were commonly categorised as confidential, and information was denied to the stakeholders. As a result, the level of objection from the stakeholder to the development plan was somewhat higher, and this situation had created conflicts within the study area (Tiwi, 1999).

Effective stakeholder participation in the coastal EIS review meetings was rarely achieved. Three of the four observed EIS review meetings were held during the implementation stage, which meant that the coastal EIS was conducted after the construction started. In these circumstances, the coastal EIS was only used as a formal procedure to fulfil requirements. Thus stakeholder participation here was not meant to provide information or input to the project plan and influence the EIS approval process (section 6.3). Another constraint on participation was the existence of elitism at the national level in the EIS approval process, as observed in the coastal EIS review meetings. At two meetings conducted within the Sea Communications office at the national level, the indications were that most of the bargaining power for decision-making was with the chairman of the meeting and eventually within the sector of government. A letter of objection from the district government concerning the case was not taken into consideration at the meetings (section 6.3). There was also domination of the technical team over other participating institutions in reviewing the coastal EIS at the meetings. These conditions will recur in the EIA approval process at national level, if the new decree on stakeholder participation is only applied to the local level. These regulations on stakeholder participation in the EIA approval process should apply to all levels, not only to the local level but also to national and provincial levels.

7.6. Lack of Empowerment of Stakeholders

Ministerial decree no. PP 51/1993 states that EIS review and approval should be conducted at national level. When EIS review meetings were held in locations in remote areas, the district government and other local stakeholders faced difficulties in attending them. Their participation at national level will require some monetary and time incentives. Anyway, the impression that their participation will only rarely influence the review process will persist. There were several limitations on stakeholders hampered their participation in the EIA process. Limitation exists within local stakeholders, since all invited district government representatives were not present at meetings, as seen in the case study. In the two meetings observed, the objection regarding the project sent by the district government to the chairman of the coastal EIS review meeting was not taken into consideration (section 6.3). This may have been a factor in the decision of the local stakeholders, especially of the district government, not to attend the coastal EIS review meetings. Another constraint was the long distance between the locations of the cases that limits stakeholder participation in the coastal EIA process (section 6.3). For projects that were approved at national level or that were national projects, a national consultant was usually hired. The distance between case locations limits the contacts or communications between the consultant preparing the coastal EIS and local stakeholders. The deadline for EIS compilation was also commonly used as an excuse by consultants for not obtaining stakeholder participation, especially for environmentally sensitive projects. The project initiators were afraid that by involving more stakeholders the time schedule for the implementation of the project plan will be disturbed. However, as one case in the project study area for a jetty development project demonstrated, not involving the local stakeholders created conflicts with the surrounding community, which led to project termination (section 3.5). The language, full in technical terms, used in the presentation of the coastal EIS was another limitation on the stakeholders in understanding the EIS, as it was not understood by stakeholders unfamiliar with the case. This made the results and findings of the coastal EIS inaccessible to the stakeholders. Since the new decree on stakeholder participation is not made public and the decree is not yet accompanied by guidance on either the implementation or on raising participation awareness. Therefore, the empowerment of stakeholders on the EIA process is more difficult. Although there was satisfactory awareness by the respondents on possible environmental impacts generated by the projects as well as on giving their input or information to the project, they were being ignored in the pre-coastal EIS compilation. Dissemination of information on the project was limited, and the participation of public on project approval was lacking (section 6.4). These would suggest that empowerment of stakeholders on participation was lacking.

The lack of empowerment of coastal communities and marine resource users as in the case study was commonly took place in other areas in Indonesia too as stated by Dahuri and Dutton (2000). They mentioned that the lack of empowerment of communities was caused by limited equity of coastal communities on coastal resource exploitation despite the considerable efforts of government to address poverty of coastal community in the 1990s. These limitations were reported among others due to limited of infrastructure, skill, technology, and capital within the coastal communities, as well as the dominance of national investor on coastal and marine management.

7.7. Future direction of stakeholder participation

New decree no. 08/2000 on stakeholder participation in the EIA process has provided the basis for improved stakeholder participation, in particular, public participation, and has provided for more active participation of the stakeholders compared to the previous law. The public can participate in almost all stages of the EIS compilation and review. In the early stage of the development plan and drafting of the terms of reference (ToR) for the EIS, stakeholder participation is passive; the developer gives information to the public and invites comments. The public then gives written comments to the government institutions dealing with the EIA at national and local levels, the highest level being the governor. In practice, however, it is difficult for the public to make objections to officials at higher ranks of government. The objection or comment should be directed to the lower level of local government such as the village office, with a copy to a higher level of institution. In the EIS review, the decree provides for active participation of the public representatives in EIS review. Besides, the public can write inputs and comments directed to the government institutions related to the EIA and the initiator, within 45 days after the meeting schedule is communicated to them. It is unlikely that the public will give or write comments beforehand if enough information concerning the plan and the EIS is not supplied. The law that gives authority for the EIS review process and approval to the local level will eventually omit the limitations on local stakeholders, as discussed before, although some limitations will still exist under the decree. The meaning of "the public" should be a broadened so that it is not be limited to the community as the public in general but also includes groups of stakeholders. The stakeholders should participate in all the EIA stages, from project conception, design and site selection, scoping of key issues and preparation of the EIS, to impact mitigation and monitoring. The implementation of the decree requires the political will of the various groups involved in the EIA process, especially the national government. The role of voluntary organisations and the public in general is to promote and raise awareness of environmental issues. Stakeholder participation in the EIA process is needed and should be supported.

Proactive and innovative mechanisms for stakeholder participation are necessary, as has been discussed by Stephenson *et al.* (1994), in case studies done in Ireland. In one case study, measures used for stakeholder participation involved consultation papers and public meetings. A consultation paper consists of a ten-page summary and a "response form" inviting written comments on the plan to compile an EIS. The form requested details of additional important topics to be added or details of topics that were considered irrelevant. Two public meetings were held to allow for the expression of views and concerns. The purpose of these meetings was not to arrive at a consensus; they were a democratic exercise and a vehicle for expression of views. All written comments and oral transcripts were reviewed to identify the relative importance of the issues and this information was used as the basis for the terms of reference of the EIS. With this method, the socio-economic nature of the concerns was analysed better and the quality of the EIS was vastly improved using stakeholder participation to broaden the scope of the EIS. In another case study, procedures for public participation included public displays to present the development plan and to invite comments before making a formal planning application and before completing the EIS.

The effectiveness of stakeholder participation in the EIA process depends on the interest and dedication of stakeholders. Most mechanisms for stakeholder participation clearly depend on the interests of different stakeholders and how these are taken into account. The limited experience on stakeholder participation in the EIA process might discourage developers from using this concept in their EIA process, if it is not accompanied by sufficient law enforcement, rewards and penalties by the government. The progress in the use of this concept has to be demonstrated to the developers by means of more examples.

7.8. Improving Use of Environmental Information and Stakeholder Participation

7.8.1. Promoting better use of environmental information

Promoting better use of environmental information in the coastal EIA process has been recognised and several ways to meet the lack of information is proposed in the publications. These are among others, as follows:

- building a strategic view of information to raise awareness and to ensure participation at all EIA stages,
- obtaining the same knowledge to understand the coastal process and its problems in order to be integrated into the management process and information exchange mechanism,
- establishment of a research agenda that has a wider user-focus and utilisation of reports on the status of the coastal environment,
- developing local or regional centres to maintain local information exchange networks and to ensure the dissemination of information (Uhel, 1999), and
- development of an information exchange network for data retrieval from one location to supplement data at another location, or to synthesise one form of data to generate another form of data (Ahmad and Sammy, 1987).

In addition, Brown (1994) proposed recommendations for improving communication and information flows among the local governments based on her research in Australia, among others, are:

- Recognition on focus of responsibility, by identification of issues on responsibilities of each policy sector for integrated environmental management, development on policy guidelines on the responsibilities of national government in relation to the local government activities, and identification of points of co-operation among the government levels.
- Establishment of a national and local information system, by the development of a strategy to co-ordinate the information sharing between national-local governments, development of an open information network connecting the available and essential information systems.
- Capacity building for local government, through the development of a comprehensive education strategy, an exchange programs with other sector of government, industry and conservation organisations, workshops on sharing of information, short course on training programs, and on-job training programs.
- Conduct action research for local/regional integrated environmental management, by preparing advice to the major national and local research funding programs, identification of priorities on existing local government programs, and re-oriented integrated information services.

For use of information in the case of tropical water resources, Gilpin (1995) identified problems, causes, and possible solutions. Problem of incomplete data for example would be managed by promoting main tropical research and monitoring, improving training and providing incentives to encourage local expertise, encouraging exchange of knowledge, data and staff, and promoting shared research between local governments. Problems on inadequate use of data would possibly be solved by better publicity and sharing of information between regions, involving planners and public in research and publishing data in simplified form to interest user. Promoting interdisciplinary research, adopting action research, and establishing model and checklists suitable for the EIA would be helpful to develop an accurate database.

7.8.2. Promoting stakeholder participation

The implementation of the law on stakeholder participation no. 08/2000 alone is not enough to improve stakeholder participation in the EIA process; therefore, the establishment of measures for stakeholder participation should accompany it. Furthermore, Ahmad and Sammy (1987) noted that the most important empowerment of stakeholders on participation is securing stakeholder participation aims to inform or educate the stakeholders, to identify problems, needs and values, to seek approaches to problem solving, to seek feedback to proposed solutions, to evaluate alternatives, and to resolve conflicts. Similarly, Abaza (2000) proposed several ways to empower stakeholders on participation. Among others, the adaptations of legal and institutional environment for more effective stakeholder participation, including: earlier notification of EIA decisions, easy access to EIA documents, background reports and data, opportunities of stakeholders to be listened, a written record of EIA decisions, and administrative or judicial procedures for the environmental review process can be tested.

There are numerous approaches and techniques, which can be used to promote stakeholder participation in the EIA process as suggested by Bisset (2000), among others, are the use of:

- Public meetings at village level. These may be held in smaller or larger communities and be implemented based on different set of rules; however, the view expressed commonly may not representative of the community.
- Advisory panels that involve of representatives of groups of communities and should also consist of groups that marginalized such as the women, poor, and indigenous people. The representatives of the community need to be carefully selected.
- Questionnaires using a set of designed questions directed to literate respondents.
- Interviews directed at selected group of communities or for individuals considered able to reflect certain group of communities. These measures require sufficient funding and skills.
- Open house and site visit to the project location accompanied by a member of an EIA team who can discuss and record the views of the visitors. Location, time of opening should be informed publicly so anybody could visit the location.

Beside the use of certain techniques, the establishment of institutional frameworks for public participation should also accompany the implementation of the law on stakeholder participation. As in the Bolivia's public participation law (OAS, 2001), which says the law has created one of the most innovative institutional frameworks for local participation that serves to organise and channel debate over development issues locally, and has a representational capacity on behalf of local community nationally. These local non-governmental institutions supervise municipal services, participate in public hearings, control government expenditures, and have standing to comment on development decisions and bring legal claims under relevant environmental and financial laws. In addition to the Bolivia's public participation law, OAS (2001) has suggested several measures for empowering stakeholder participation using the local non-governmental institutions, including:

- education for sustainable development directed to relevant sectors of community to improve the quality of public participation,
- training and support to build the capacity of the community especially on funding,
- building community awareness to direct attitude toward sustainability,
- improving communication to facilitate efforts at education, training and awareness building through removing barriers of language, lack of means of communication with remote areas,
- building trust at national-local through co-ordinating local decisions and linking them to national government,
- remove conflicting and overlapping jurisdictions, and
- integrate local communities into decision-making under other key development laws.

7.8.3. Promoting effective EIA, decentralisation and building local capacity

Theoretically it is accepted that EIA can be an important tool of ICZM for giving effect to sustainable development. In practice, the use of EIA will depend on the role, scope, and integrity of the process; the use of environment, economic policy and planning instruments for decision making; and the policy commitment to sustainable development (Sadler, 1996). Furthermore, Sadler proposed basic steps for sharpening the use of the EIA for sustainability instrument, including:

- taking into account the consideration bio-diversity on the use of sustainability concepts and principles in EIA,
- involving relevant criteria and indicators into screening, significance, and other EIA checklists,
- refocusing EIA on the environmental basic principles, for example on carrying capacity and polluters pay principles, and
- specifying requirements for in-kind compensation for residual impacts.

In addition, five elements for promoting EIA in developing countries are proposed by Ahmad and Sammy (1987), including:

- Political commitment to leadership, involving a willingness absolutely to use the EIA as part of the decision making process supported by the necessary national leaderships.
- Local control of EIA, in which the groups that will use the EIA result, should be the ones who guide its content and direction, incorporating local values and the management approach to control the EIA.

- Effective scoping on identification of the most important impacts of the proposed project for more practical and more cost effective approaches to EIA.
- Development of an environmental database leading to the formation of a network of all local / national data banks to avoid duplication of efforts and costs. This network would facilitate free information exchange to improve the quality of data for the EIA at low cost.
- Information sharing which involves development of an environmental data bank and forming a network of all local-national data banks to avoid duplication of efforts and costs. This network would facilitate the free information exchange to improve the quality of data for the EIA and to avoid costs.
- Training involving short-term solution of training for general managers rather than specialists on the awareness of environmental impact assessment. This approach would permit the use of locally controlled EIA as discussed before.

Decentralisation promoted intensively in Indonesia recently aims to smoothen the handling of power from the national to the district levels of government. Successful decentralisation requires some degree of local participation, while the process itself improves the opportunities for participation by giving more power and resources to the local government. There is a debate about the capacity of local government and communities to plan, finance and manage their new responsibilities. The traditional approach to decentralisation has been to build capacity before transferring responsibilities or revenues, due to worries on irresponsible spending, local corruption, regional inequalities, and the reluctance of central government to decentralise their authority (FAO, 2001). It is also noted that developing countries like Indonesia show evidence that the capacities of local government levels increase slightly but constantly in the years following decentralisation. Similarly, OAS (2001) indicated that the Bolivia's experience (section 7.8.2) shows that the decentralisation has increased local capacity and consequently the local government in gaining access to local and national resources.

The best way to build local capacity is through the process of decentralisation itself. FAO (2001) suggests that the best way to promote decentralisation, at the same time to improve local capacity would be through demand-driven capacity-building programs. One way to improve local capacity is to allow local institutions to use a partial program fund to hire the technical expertise for assistance. Furthermore, FAO (2001) proposed that based on Bolivia's experience, this program if found locally, would be quicker and more cost-effective to improve local capacity through learning by doing. Having strong local participation through non-organisational institutions, as in Bolivia it deepens the decentralisation process and motivates local government to actively improve their capacity. To promote the decentralisation, Mera (2000) proposed several alternative frequently used modes of providing capacity building in general, such as engage in academic degree and non-degree programs, engage in short-term training programs, training of trainers, on-the-job training, and consultant services.

8. RECOMMENDATIONS TO IMPROVE THE COASTAL EIA PROCESS

8.1. Introduction

This chapter first presents recommendations to improve the coastal EIA process that covers improved use of environmental information and stakeholder participation (section 8.2). Then instruments to improve the coastal EIA process that consists of Banten Bay Information System, EIS review criteria, and measures to improve stakeholder participation (section 8.3).

8.2. Improved Use of Environmental Information and Stakeholder Participation

The research results concerning the use of environmental information and stakeholder participation in coastal EIA were described in chapters 5 and 6. Based on these results, this chapter presents recommendations for:

- improved use of environmental information,
- improvement of stakeholder participation, and
- improvement of factors that are outside the scope of this research.

For immediate and cost-effective results, the coastal EIA effectiveness could be strengthened by the improvement of the use of environmental information and stakeholder participation in the area of scoping, evaluation of significance, review of coastal EIA reports, and monitoring and follow up of coastal EIA process.

8.2.1. Improved use of environmental information

Use of environmental information, in a broader sense, refers to how information is obtained and used, which also covers the need, availability, and accessibility of information sources. The recommendations for improved use of environmental information in the coastal EIA process covers the legislation basis, improvement of coastal monitoring, development of a coastal information system, and capacity building. Figure 8.1 shows the framework for improved use of environmental information in the coastal EIA process. A summary of the proposed recommendations is given in table 8.1.

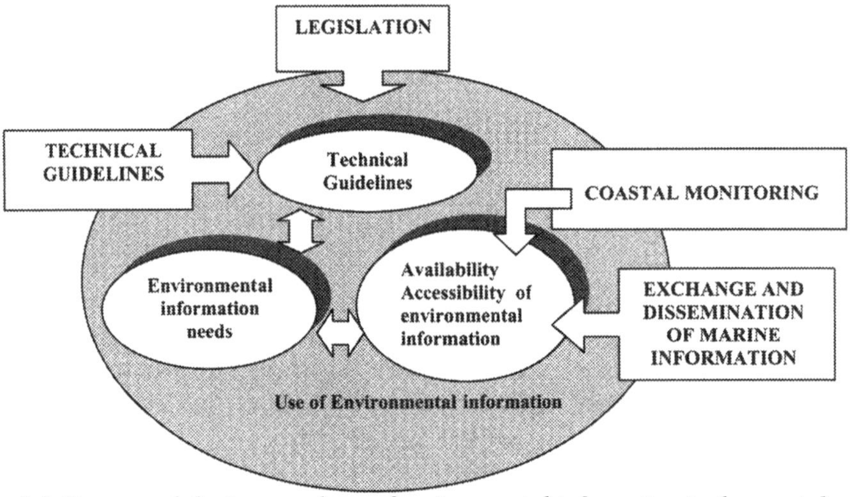

Figure 8.1. Framework for improved use of environmental information in the coastal EIA process.

Improvement of legislation

The research identified that the district government has limited access to environmental information, especially marine information of its area, since most of the information is within government institutions at national level, in research institutions or in consultants' offices (section 5.5). There is only limited use of environmental information, especially marine information, in the coastal EIA process as concluded in section 5.9.1. The district government faces difficulties in accessing that information because there are no regulations on the ownership of the district data. Therefore, to acquire the needed environmental information, the district government is suggested to regulate and provide guidance on:

- The review of relevant district regulation on coastal development and the ownership of district-level data, as this will give power to the district government to access district data available in outside offices.
- The development of a database on coastal EIA regulations and experiences concerning the application of those regulations to prevent duplication and to communicate between EIA district offices on coastal and marine issues.
- Awareness programmes on coastal and marine sustainability and information dissemination for the educated people, including the government staff, the managers, and the university students.
- Review areas of responsibility between district offices for marine information contribution and coastal monitoring.
- The creation of a marine information exchange system at district level for data retrieval and data synthesis, with the help of the national government and research institutions.
- Translation of the available marine data on international and national research reports into local information.

- Promotion of research on coastal and marine resources in collaboration with research institution using the ongoing research programmes within the Ministry for Research and Technology.

It is suggested that the national government regulates and provides guidance on:

- The foster of the flow of marine information that will allow the sharing and dissemination of marine information between and within government offices at all levels.
- Giving responsibilities on developers and project proponents to monitor significant environmental impacts and linking impact management and coastal and marine monitoring as part of coastal EIA process.
- The supports including the funding system and incentives as well as the ways to allow and encourage other stakeholders in the integration of marine information and to seek the effective partnership to engage in the provision of marine information.
- The use of a standard EIS review criteria in the coastal EIS review process.

Improvement of EIA technical guidelines
As discussed in section 5.9.2, the marine information stipulated in the EIA technical guidelines is limited. It is suggested that:

- The national government should review these EIA guidelines regularly and adjust or revise them to match the current status of EIA practice.
- The separation of the technical guidelines for major coastal projects and those for minor coastal projects, and development of an EIA technical guideline that focuses on coastal development.
- The addition of indicators and parameters for biology component in particular marine biology components to the technical guidelines.
- Stipulate and control the conduct of impact monitoring by the project owner.
- Enforcement and control the distribution of coastal EIS reports to the relevant stakeholder groups before review meetings.

Improvement of coastal monitoring at district level
The limited use of information sources, in particular marine information sources, to compose and review the coastal EIS in the study area is among others due to limited district coastal monitoring activities (section 5.9.2). It is suggested that:

- The existing coastal monitoring activities should be improved and co-ordinated by the district government.
- Promoting shared coastal monitoring programmes between adjacent district governments in the coastal region, utilising common monitoring indicators.
- A coastal monitoring system that involves all levels of government below the district level should be developed.
- Stakeholders at sub-district, village, and sub-village levels should be active in coastal monitoring, empower coastal community groups such as fishermen.

farmers, and fishing boat owners in coastal monitoring and used as an early warning system for selected coastal environmental indicators.

The national government is suggested to:

- Stipulate common indicators and parameters for coastal and marine monitoring.
- Provide support facilities, capacity, and funding for district coastal monitoring.

Improvement of exchange and dissemination of marine information

To support the improved use of environmental information in the coastal EIA process, priority should be given to the exchange and dissemination of the available coastal and marine information. Several ways are suggested as follows:

- The marine information of the district area that exists in the coastal EIS and in research reports should be reviewed; information related to basic data of the marine area should be extracted and compiled.
- The development of a meta-data system as the most simple way and useful tool to allow the district governments to find out information on what data are available, where they are kept, and who are the holders.
- Development of a strategy to co-ordinate the information sharing and open information networks among the available information systems.
- District governments in the coastal region in collaboration with research institutes and universities should develop a coastal information system.
- Improvement of and support for sharing coastal information among district offices should be promoted and fostered.
- Linking mechanisms between national and local networks should be developed through the establishment of local information networks, and accessibility to computer networks among district government offices should be improved.
- Raising awareness of the professionals, the school children and university students to improve the integration of information, to transfer the knowledge, and to disseminate information.
- Provide supports and the funding system and the mechanism to encourage stakeholders and to find an effective partnership for the provision of marine information.

Capacity building

The capacity of the district government to use environmental information in the coastal EIA process is limited (section 5.9.2). Therefore, it is suggested that district governments should take steps:

- To enhance the district government's awareness and understanding of the coastal EIA process, coastal process, and sustainability of coastal and marine areas.
- To develop a comprehensive education strategy for district government staffs among others through exchange programs with sector of government and industry, and on-job training programs.

- To improve the institutional capacity among others by engage in an action coastal and marine research conducted in collaboration with coastal research institutions, universities, and NGOs.
- To improve access and district expertise on EIS review process.

As the capacity of the district government to conduct coastal monitoring activities is limited (section 5.7), the national government should support the district government by:

- Improving laboratory facilities and human resources for the development of district coastal monitoring.
- Raising awareness of all stakeholders on the need for coastal monitoring.
- Initiate training for trainers on simple coastal monitoring activities at district level.
- Provide training on methodology to obtain and analyse data to the staff of district government responsible for environmental monitoring.
- Development of standard procedures for the daily coastal monitoring of indicators and parameters to determine common significant impacts.

Concerning the coastal EIS approval, it is also suggested that the national government should:

- Assist the district government to improve its capacity to manage coastal EIS approval.
- Develop review criteria and scoring procedures in order to standardise and simplify the EIS review process by the district EIA commission.
- Assess and improve the expertise available to the coastal EIS review process at district level to facilitate the carrying out of review activities with the assistance of the central EIA commission research institutions, universities and NGOs.
- Provide training to improve knowledge and methodology concerning the coastal EIA to the district EIA commission.
- Promote the exchange of marine information and experiences concerning the coastal EIS review among district governments within the region.

Table 8.1 presents recommendations for the improved use of environmental information in coastal EIA process.

Table 8.1. Recommendations for the improved use of environmental information in the coastal EIA process.

Improvement of legislation	
District government	- Review relevant district legislation on coastal development, develop data-base on EIA regulations and experiences, review areas of responsibility between district offices for marine information contribution - Regulate access to district marine information and campaign of awareness on information dissemination, translation of available data on research reports, and creation of a marine information exchange system - Promote research activities on marine resources in collaboration with research institutions at national level using the ongoing research programmes
National government	- Foster information sharing and information dissemination - Stipulate responsibilities on impact monitoring for project proponents - Conduct of more coastal and marine research using ongoing research programme, provide support on incentives and ways for provision of information - Stipulate use of EIS review criteria for EIS review process.
Improvement of EIA technical guidelines	
National government	- Review EIA regulation, provide guidelines for major and minor coastal projects, add marine biology indicators to the EIA technical guidelines - Redefine permit for major and minor coastal activities; provide guidance and procedures for permit approval - Stipulate and control the conduct of impact monitoring by the project owner - Enforce and control the distribution of EIA reports
Improvement of coastal monitoring at district level	
District government	- Identify important coastal monitoring indicators, empower community for monitoring - Coordinate coastal and marine monitoring, develop a coastal monitoring system for all levels of district government - Promote shared monitoring programmes, utilise common monitoring indicators
National government	- Stipulate indicators and parameters for coastal monitoring - Provide support facilities, capacity, and funding for district coastal monitoring
Improvement of exchange and dissemination of marine information	
District government	- Provide mechanisms to promote marine information exchange - Provide marine information in collaboration with research bodies - Develop a metadata system and a coastal information system at district level based on research data, develop local information exchange networks - Develop a strategy to coordinate information sharing and raising awareness - Provide support, funding and mechanism for encouragement on information exchange
Capacity building	
District government	- Improve awareness of coastal processes and the EIA process - Provide training for trainers to improve coastal monitoring - Assess and improve expertise in the EIS review process, improve capacity through exchange of expertise and on job training, engage in action research program in coastal and marine research.
National government	- Develop review criteria and scoring procedure on EIS review - Improve district capacity for managing coastal permit system and methodology for coastal monitoring, and raise awareness - Improve capacity and methodology of district EIA commission - Promote exchange of marine information and expertise on EIS review

8.2.2. Improvement of stakeholder participation

The recommendations to improve stakeholder participation in the coastal EIA process are improvement of the legislation basis, capacity building, and development of measures for stakeholder participation. Figure 8.2 shows the framework for the improvement of stakeholder participation in the coastal EIA process. A summary of the recommendations is given in table 8.2.

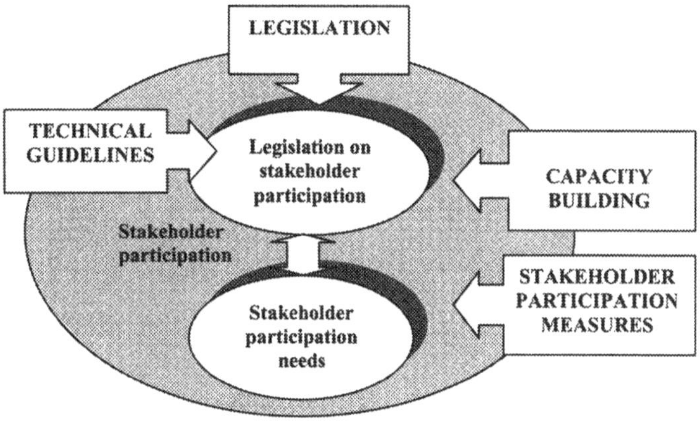

Figure 8.2. Framework for the improvement of stakeholder participation in the coastal EIA process.

Improvement of legislation

There is limited stakeholder participation in the coastal EIA process, as was concluded in section 6.5.2. Effective stakeholder participation in the coastal EIA process requires the inclusion of basic principles of stakeholder participation in the process as discussed in section 2.3. Improvement of legislation is suggested to the national and district governments, among others, through:

- The making of district regulations on stakeholder participation as a follow-up to decree no. 08/2000 on stakeholder participation and access to information in the coastal EIA process.
- The creation of legal and institutional arrangements for more effective participation, including easy access to coastal EIS, reports and data; and chances of stakeholders to provide input and information.
- The provision of practical and innovative guidance for the ways in which various stakeholders can communicate and negotiate in addressing environmental issues.
- Development of practical and innovative guidance for empowerment of different local stakeholder groups, such as groups of woman, youth, farmers and fishermen, and stakeholders from the private sector and business.
- The creation of local non-government institution through the improvement of function of available local forum at district level to enhance participation.
- Provide incentives and rewards for industries that utilise stakeholder participation in their EIA process.

Improvement of technical guidelines
Since the decree on stakeholder participation in the coastal EIA process is relatively new in Indonesia, knowledge about it needs to be disseminated more widely within every district government (section 6.5.2). Several ways are suggested as follows:

- Guidelines to increase participation awareness among policy makers, local staffs and the general public should be developed by the national government.
- Development of different types of guidance to assist practitioners in communication with various stakeholder groups.
- The improved guidelines should also include the requirement for a written agreement from local stakeholder groups for the approval of the coastal EIS.
- Proactive support from national legislation in the form of incentives and awards should be stipulated to enhance the implementation of stakeholder participation.
- Specific mechanisms should be provided, in detail, the requirements of public notification, public consultation, public hearings, advisory groups, and responsiveness summaries on stakeholder participation in the coastal EIS review process.

Capacity building
The community has limited experience and a limited knowledge base on stakeholder participation, as was concluded in section 6.5.1. It is suggested for the district government that they should:

- Provide education to the coastal community for the improvement of participation awareness, as it will ensure a more effective participatory process.
- Take an active role in promoting participation awareness through education and training, in collaboration with public and private institutions as well as non-governmental organisations.
- Develop training courses and teaching materials and adapt it for target groups, especially women, youth, farmers, and fishermen, using local languages and illustrations to convey the message clearly.
- Educate the public about the link between stakeholder participation in coastal EIA and the coastal environment and its importance should be carried out through educational institutions, beginning at primary school level.
- Improve their institutional capacity to implement the concept of stakeholder participation, promote exchange of experiences with other district offices.
- Establish a district consultative group consisting of representatives of coastal community groups that can be used as a mediator among the coastal community, between the community and district government and developers. This group could give input on stakeholder participation, supervise the participation in public hearings, and become a group having standing to comment on development decisions.

It is suggested that the national government should support the capacity building of district governments by:

- Providing mechanisms for the participation of a wide segment of stakeholders throughout the coastal EIA process.
- Promoting co-operation between district governments within a coastal region to enhance the implementation of the concept of stakeholder participation by district governments
- Promote the exchange of experiences on stakeholder participation among governments at various levels.
- Development of a national consultation group in order to ensure that the concept of stakeholder participation is implemented and to ensure a routine dialogue between governments and other stakeholders.

Development of measures for empowerment of stakeholders

Some measures available for stakeholder participation and information dissemination are defined by the State Ministerial Decree of Environment no. 08/2000. However, the available measures are limited to participation in the coastal EIS review process, while measures for empowering stakeholders are not available (section 6.5). Therefore, the district governments are suggested to:

- Encourage active participation of local universities and research institutions.
- Support the development of a body of local stakeholder groups in each district area, as the lowest level of authority for the coastal EIS approval. It will determine which groups of stakeholders are really affected by the project and which stakeholder groups have strong influence on the project.
- Improvement of means of communication with community in remote areas in order to improve distribution of information builds up community awareness.
- Improvement of quality of participation through informal education to all relevant sectors of community,
- Provide support, funding and mechanism for empowering stakeholder participation among others through co-operation with research institutions.

Table 8.2 presents recommendations for the improvement of stakeholder participation.

Table 8.2. Recommendations for the improvement of stakeholder participation in the coastal EIA process.

Improvement of legislation	
District government	- Develop legal and institutional mechanisms for empowerment of stakeholders, for broader participation, dissemination of the decree no. 08/2000, and provide support for the implementation of the decree.
National government	- Provide specific mechanisms for participation in the EIS review process, and to make a written agreement from the community groups a requirement for EIS approval. - Provide incentive and rewards for carrying out stakeholder participation in EIA process. - Improve and utilise local forums at village level to improve participation.
Improvement of technical guidelines	
National government	- Provide guideline to increase participation awareness on coastal environment among district governments; and to support the implementation of the decree on stakeholder participation. - Provide different types of guidance and mechanisms for empowering stakeholders. - Proactive support with providing incentives and award to improve participation.
Capacity building	
District government	- Improve participation awareness through education and training for target community groups and community in general. - Improve institutional capacity to implement the decree on stakeholder participation, promote co-operation and exchange of experiences with other district governments. - Provide practical assistance to improve participation and awareness, invite input, and disseminate information. - Establish a district consultative group to assist participation in general.
National government	- More dissemination of the concept; provide technical assistance and training to improve participation. - Provide support for co-operation among district governments; promote exchange of expertise and experiences among various government levels. - Develop a national consultation group for concept implementation.
Empowerment of stakeholders	
District government	- Encourage active participation of local universities and research institutions - Develop measures to improve awareness, to empower stakeholder, to identify stakeholder, to disseminate information, and to invite input - Improve of means of communication among stakeholders, provide support, funding and mechanisms for empowering stakeholder. - Use of informal education to improve quality of participation of all relevant sectors of the community - Support the development of a local body of stakeholder groups for improvement of stakeholder participation.

8.2.3. Improvement of external factors

The implementation of the proposed recommendations and measures for improving use of environmental information and stakeholder participation by district government will not automatically bring about an effective use of the coastal EIA process. And the effective use of the coastal EIA process will not automatically improve ICZM; it will only partly contribute to its improvement. There are external factors that should be improved in order for the district government to use the coastal EIA effectively.

- Because the sectoral offices at national level still manage relatively small-scale projects of coastal related activities (section 3.3.3), it is suggested that the national government delegate the authority to issue some permits to the district level, in the spirit of decentralisation.
- The permit systems should also be redefined; permits for major coastal activities should be distinguished from those for minor activities. For example, authority to issue permits for major developments such as harbour development and large-scale reclamation should stay at the national level, while that for minor coastal projects, for example jetty development, sea sand mining and small dredging activities, should be decentralised to the level of district government.
- Stronger collaboration between district governments in the coastal region should be promoted, better communication between and within them should be enhanced, and co-ordination in coastal programs to achieve better coastal protection and avoid duplication should be supported.
- Local empowerment should be improved and effective district governance should be promoted to make the district government more accountable.
- Research collaboration with research institution at national level should be promoted, in particular on the provision of local information and stakeholder participation in the coastal EIA process using the ongoing research programme on dissemination of technology.

The implementation of the decentralisation process should be pursued, consistently developed and revised, as well as supported by the national government on:

- Rules on the decentralisation should be more explicit and reasonably permanent to reduce uncertainty within the district government.
- The enhancement of technical guidelines and technical assistance for various aspects of decentralisation and improvement of district institutional capacity.
- The creation of capacity building programs through technical assistance and learning by doing based on the needs to promote decentralisation process.
- The promotion of decentralisation of policies within district government and the incorporation of district interest into the national interest.
- The inclusion of pressure from outside groups such as national and international NGOs, other stakeholders, research institutions and universities to make the implementation of the decentralisation process consistent at national and district levels.

Figure 8.3. presents framework for improved use of environmental information and stakeholder participation in the coastal EIA process.

Sustainable development of coastal areas

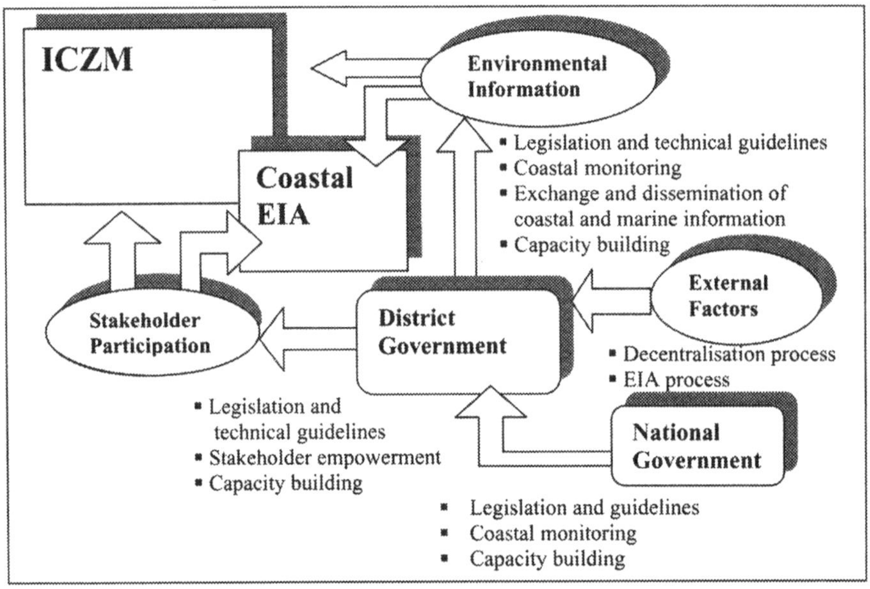

Figure 8.3. Framework for improved use of environmental information and stakeholder participation in the coastal EIA process.

8.3. Instruments to Implement Some of the Recommendations

Based on the recommendations proposed in the previous section, three measures have been developed to improve the use of environmental information and to improve stakeholder participation in the coastal EIA process. They are the Banten Bay Information System (BBIS) developed by the Teluk Banten Research Programme, and two other measures (EIS review criteria and measures to improve stakeholder participation) developed in the context of this thesis.

8.3.1. Banten Bay Information System (BBIS)

Why the BBIS was developed

The Banten Bay Information System (BBIS) was developed to improve better understanding and awareness of the coastal process and its interactions with human activities, through the integration of research information generated by the Teluk Banten Research Programme. The development of the BBIS took place during the past 3 years (1999-2001) and three versions of the system evolved during this period. Its development involved continued consultations with the District Planning Office (*Bappeda*) and the District Environmental Office of the District of Serang, which were selected as the end users of the system given their roles in the coastal EIA process. The last version of the BBIS was distributed to district governments of the study area and participants of the final Teluk Banten Research Programme workshops in Cilegon-Banten province (25-26 September 2001) and in Jakarta (27 September 2001).

Information in the BBIS
The BBIS consists of seven information modules:

1. The *Introduction Module* gives introductory information about the Banten Bay, its natural environment, main human activities and related issues.
2. The *Stakeholder Impact Module* provides a list of stakeholder groups for the main activities in different regions of the Banten Bay, expected impacts of the activities in the marine environment, and a list of information available with stakeholders.
3. The *Marine Environmental Module* gives a description of the a biotic and biotic status of the four important marine ecosystem in the Teluk Banten Research Programme (coral reef, sea grass beds, grouper and snapper fishery, and water birds of Pulau Dua).
4. The *Environmental Function Module* gives an overview of the environmental functions of the Banten Bay ecosystem and related stakeholders.
5. The *Coastal Zone Management Module* gives suggestions on how the information stored in the previous modules can be used to support coastal zone management.
6. The *EIA Module* provides information on the EIA stages, links to related information provided in the BBIS, the EIS review criteria to simplify the EIS review process and a list of Indonesian regulations.
7. The *Coastal Zone Monitoring Module* gives suggestions for monitoring the Banten Bay coastal zone.

Quick tour on the BBIS is presented in Appendix VII.

Implications of the use of the BBIS by district government
During in-depth interviews with representative of the district government (28 September 2001) when the system was expounded to them, they indicated that the information that could be provided by the BBIS is very much needed by district government. In particular, they mentioned the baseline information on four marine ecosystems that is not yet available. The system is useful to district government to support the exchange and dissemination of marine and coastal information in particular for the EIA process. It can be developed further to set up an integrated coastal information system to support coastal planning and monitoring activities using the available research program on dissemination of technology. Improvements that could be suggested to the BBIS system include the translation of the English language used to Indonesian and the inclusion of information on indicators needed for coastal monitoring. Use of indicators on four main ecosystems under studied for coastal monitoring should be disseminated to the community in general and to the school children in particular to raise awareness on coastal sustainability. The researchers involved in the development of the system should conduct awareness programmes and training to the district governments in the adjacent coastal areas on the use of the system. Experiences gained on the development and use of the system should be disseminated to other district government in the region with similar coastal ecosystems.

8.3.2. EIS review criteria

Proposed criteria

The proposed EIS review criteria are developed by updating the EIS review criteria used in the EU countries (Lee *et. al.*, 1999) and adjusting them to the State Ministerial Decree of Environment no. 02/2000 on guidance for the EIS review. The proposed EIS review criteria include a list of review topics, review and scoring procedures and a collation sheet. The scoring procedure is developed by this research study for further quality assessment of the original review criteria. It consists of steps to quantify the assessment quality of the EIS reviewed. The review criteria are explained in detail in section 4.4.3 where methodology for assessment of the EIS quality is discussed. The review criteria have been used in this research and have proven to simplify the review procedure, and they were considered as more objective because there is the application of a scoring procedure (section 5.3). A tryout of the review criteria by district government resulted in a recommendation that the district government also needs a list of review criteria to assess the terms of reference (ToR) of the EIS. Based on that information, such a list of review criteria is developed to accompany the EIS review criteria. For further information on the description of these review criteria see Appendixes III and IV.

To evaluate the proposed review criteria, the district government tested the measures prescribed in the criteria for use in the actual review process. A meeting of 17 members of the district EIS review committee, representing 15 district offices of the study area (the District of Serang), was held on 29 September 2001 to explain and discuss the use of the measures. The participants were requested to answer a questionnaire to evaluate the measures, which sought information on the benefits of the use of the measures and constraints experienced, and requested comments on their use and improvement. The questionnaire was answered and returned by 10 of the 17 participants. The district review committee members responded that they need these review criteria, which were previously not available, to help them in the EIS review process. Moreover, they said that the proposed measures had more detailed review criteria; they were more practical, simple and easily understood, more structured and more comprehensive compared to the available guidance. Two of the review committee members indicated that in order to have more sensitive results, the use of the scoring procedure for an EIS should be accompanied by weighted ranks on the importance of each environmental component based on impacts of the proposed plan. As for this comment, since the review criteria are mostly general, the use of weighted ranks and minimal quality for the EIS to be approved should be decided by the committee members before the review process is conducted.

Achieving more effective review results

The district government needs assistance to conduct the coastal EIS review process. The assistance could come from research institutions or universities for a limited time, for example 1-2 years, of the coastal EIS review process. Through a process of learning by doing, the district governments are expected to improve their capability to conduct the coastal EIS review process. In order to have more effective review results, it is suggested that the review process be preceded by a visit to the location of the planned development, so that the committee members will have more comprehensive knowledge regarding impact analysis within the EIS review.

8.3.3. Measures for empowering stakeholders

Here measures are proposed to identify stakeholders and to improve participation. These measures are based on the research results, inputs from district government and relevant knowledge and experience gathered through literature reviews.

Measure to identify stakeholders

In the Banten Bay coastal zone, 5 types of development area are distinguished in which 19 types of human activities are involved. The main activities within each development areas are presented in table 8.3.

Table 8.3. Main activities according to land use designation in the Banten Bay area.

Banten Bay coastal zone	***Main activities***
1. Bay area	(1) fishing involving various techniques (2) coral mining (3) grouper fishery (4) coral collection (5) bay traffic (6) seaweed culture (7) sea sand mining
2. Land area mainly for industrial development	(8) industrial development (9) harbour and jetty development (10) land reclamation
3. Land area for multiple-use development	(11) urban development (12) agriculture (13) harbour (14) fish auction (15) wood industry (16) P. Dua bird sanctuary island (17) multiple-use urban and industrial development
4. Land area mainly for agriculture	(12) agriculture (18) aquaculture
5. Catchment area of river	(17) multiple-use urban and industrial development (19) multiple-use urban development and agriculture

Source: The Banten Bay Information System.

There are potential stakeholder groups associated with these activities. The stakeholder groups can be distinguished as groups from government offices at district, provincial and national levels and groups of stakeholders within the community as well as within the private sector. Some groups are also associated with research institutions at various levels related to the main activities. Based on a list of the identified stakeholder groups described before, a more detailed list has been elaborated. An example of main stakeholders for harbour and jetty development in the Banten Bay area is given in table 8.4.

Table 8.4. Main stakeholders for harbour and jetty development in the Banten Bay.

Groups of stakeholders	***Stakeholders***
District government:	District Planning Office/*Bappeda,* Sub-district offices, Village offices, District Environment office, District Mining offices, District Fishery office, Banten harbour authority
Provincial government	None
National government	Ministry of Communications, *Dyshidros*, State Ministry of Environment/ *Bapedal*
Citizen / Community group	Fishermen, middlemen, fishermen groups, farmers, Indonesian fishermen groups, rural community in general

Source: Banten Bay Information System.

Measures to empower stakeholder

There are local forums available at village level, which could be used to improve participation awareness and to convey information. Two local forums organised by the head of village offices to meet with their communities are the *sepasaran*, a forum held every 5 days, and the *selapanan,* held every 35 days. Other regular informal forums as the *pengajian* (religious meetings), *arisan* (informal community meetings), and *posyandu* (a forum used as a facility to check the health of children and pregnant women) can be used to increase participation awareness on the sustainability of the marine environment. Practical guidelines and procedures for raising participation awareness written in the local language should accompany the use of the local forums. It is suggested that in order to get a broader audience, the district government in co-operation with the national government should use the available governmental programs in mass media such as radio and television to carry messages on participation awareness. The available programs are *Siaran Pedesaan* (Information from villages), *Kontak Tani dan Nelayan* (Contacts among farmers and fishermen), and *Selamatkan Karang Sekarang* (Save coral reef now). Other programs on radio and television that can be used to improve participation awareness are *sinetron* (local soap operas), *wayang kulit* (puppet shows), *ludruk* (humour stories and shows), and *cerita rakyat* (folk shows). Some international researchers, for example Smith (1993), have suggested measures to obtain inputs from stakeholders. However, not all of them are applicable to local stakeholders with limited experience on stakeholder participation, as in the Banten Bay area. The appropriate measures to invite inputs from the stakeholders or the public are survey techniques using interviews and questionnaires, as well as the use of available local forums. For interviews, local guides who can translate questionnaires into the local language should accompany the survey teams.

Implementation of the measures by the district government

Funding available for district government offices to improve their institutional capacity, to improve stakeholder participation and to provide training and courses is limited. Therefore, it is suggested that the district government collaborate with the national government and national and international organisations to develop alternative and flexible funding mechanisms to promote stakeholder participation.

To improve participation in EIS review process, the stakeholders should be supported by sufficient related information so that they could provide input or comments. Sufficient copies of the ToR and the EIS, and other supporting materials should be distributed to the District EIA Commission and related stakeholders groups before the EIS review meeting. A summary of the development plan, the impacts generated and the measures for impact management, written in simple language, should be provided to the target community groups. Public hearings using local forums should be conducted along with the distribution of written materials. Enough time should be provided to the stakeholders to review these documents and provide comments. Instructions on providing inputs and comments should be provided using simple language.

8.4. Conclusions

As discussed in chapters 5, 6, and 7, the use of environmental information and stakeholder participation in the coastal EIA process are not balanced. In order to improve the effectiveness of the coastal EIA process, recommendations for the improved use of environmental information and stakeholder participation in the coastal EIA as well as improvement of factors outside the coastal EIA process have been proposed. Three instruments have been developed to support the implementation of some proposed recommendation by the district government. Evaluation of the three instruments has been conducted in order to improve the effective use of these instruments.

PART IV

CONCLUSION

9.
CONCLUSIONS

9.1. Introduction

This research explored the ways in which the Environmental Impact Assessment (EIA) can be more effectively utilised in development activities so that the environment damage could be minimised. It also explored weak areas of coastal EIA and made recommendations for rectification coastal EIA to improve integrated coastal zone management (ICZM). The conclusions from this research comprise of areflection to the central problem statements (section 9.2.); to the conceptual and application relevance of the research (section 9.3); to the methodologies of the research, research design, study cases, and research findings (section 9.4); and suggestions for further research (section 9.5).

9.2. Reflection on Central Problem Statements

How can environmental information and stakeholder participation contribute to more effective use of coastal EIA for ICZM?
Based on a thorough analysis of the research on the use of environmental information and stakeholder participation in the coastal EIA process (chapter 5-6-7), improvements are suggested (chapter 8). These improvements are suggested for legislation, the EIA guidelines, the Environmental Impact Statement (EIS) review process, coastal monitoring, stakeholder participation, and to institutional capacity building. These improvements are achieved by:

- strengthening the use of marine information through the addition of marine indicators in the EIA guidelines and improvement of the exchange and dissemination of marine information,
- simplifying the coastal EIS review process through the use of review criteria,
- improving stakeholder participation through the provision of a clear and specific role of the stakeholder, and
- using of measures for participation in the coastal EIS review and in the development approval processes.

How can improved use of environmental information and stakeholder participation be utilised by the district government for maximum benefit?
More effective use of coastal EIA has been suggested through the improved use of environmental information and improvement of stakeholder participation in the coastal EIA process, and to use EIA review criteria to improve coastal EIS review process. These suggestions have been embedded into the Banten Bay Information System (BBIS). To enable the district government to use the BBIS more effectively, improvements are suggested to the BBIS on:

- the translation of information from the original language, English, into the Indonesian language, and
- training for trainers and awareness programmes on for how to use of the system to the staff of district government and of local universities.

9.3. Reflection on conceptualisation and application

Conceptualisation

The use of marine environmental information and stakeholder participation has the potential to improve coastal EIA as an effective coastal management tool and to improve ICZM.

The research conducted for this thesis concluded that coastal EIA indeed can be used cost-effectively as a management tool to improve ICZM. Through improvement of use of environmental information and stakeholder participation in legislation and technical guidelines, institutional capacity, stakeholder participation, exchange and dissemination of coastal and marine information, and coastal monitoring, the district government will have an increased support for analysing and approving the EIS for coastal development, and thus the quality of the coastal EIA will be improved significantly. This will improve the quality of information on which coastal developments are approved and better quality of the decision on coastal zone development projects.

Environmental information generated by an integrated research team will contribute to better management decisions in ICZM.

In order to better understand a complex coastal system, the natural as well as human dimension should be studied in order to provide information needed by coastal managers. Banten Bay Information System (BBIS) is a prototype information system that aims to better understand the coastal processes and to assist the district government of the Banten area in conducting EIA and ICZM. The system contains information from the Teluk Banten Research Programme and was developed in close collaboration with the district government. BBIS provides the district government with integrated environmental information on selected coastal and marine aspects on sea grass, fishery, coral reef, and birds as selected parameters for environmental monitoring. By making these results of integrated research accessible to coastal managers, BBIS will support marine and coastal management conducted by the district government of the Banten area.

The changes of policies within the new Indonesian government have the potential to improve the role of district government in coastal EIA and ICZM.

The changes in Indonesian policies particularly in the decentralisation process in 1999 influence the implementation of acts and policies such as regulation and decrees related to the EIA. Decrees passed on the establishment of procedures to review EIS, on the formation of district EIA commission and on stakeholder participation in the EIA process have significantly improved the role of district government in coastal EIS approval. The establishment of Banten province in 2000, where the Banten Bay is located, significantly increased the power of district government to manage their marine and coastal area and directly to approve coastal development projects. This more active role of district government in ICZM increases the need for effective coastal EIA at the district level. The results of the study suggest that through improved use of environmental information and improvement of stakeholder participation in the coastal EIA, the district government will have better support to use coastal EIA effectively to improve ICZM. The establishment of the Ministry of Marine and Fishery promotes coastal and marine management and increases local autonomy and participatory of the district government in promoting ICZM.

Application

The recommendations proposed to the national government apply, in particular, to the State Ministry of Environment, the Ministry of Communications, and to the Ministry of Marine and Fishery as well as to the District Government of Serang. Additionally the following suggestions are made to the Ministry of Marine and Fishery:

- to develop legislation on Integrated Coastal Zone Management as an umbrella for integrated sectoral strategies on coastal and marine management as well as legislation on capacity building including on financial assistance,
- to initiate and regulate public private partnership on sustainable coastal and marine resource management, and
- to stipulate the use of coastal EIA as a management tool for ICZM.

The following suggestions are made to the District Government of Serang:

- to merge the functions of the District Planning Office and the District Environment Office in order to avoid duplication and contradiction on policies related to coastal and marine resources management,
- to set up a District Research Office with duty to coordinate and improve local research agenda, and
- to use the ongoing research programme on technology dissemination of the Ministry of Research and Technology to improve the provision of information on marine biology and awareness on stakeholder participation.

9.4. Discussion of the Research

Methodology

Assessments of the use of environmental information in the coastal EIA process was conducted through 6 activities:

1. Assessment of environmental information stated in 3 EIA guidelines,
2. Assessment of environmental information described in 10 coastal EISs,
3. Assessment of environmental information sources to draft and review EISs involving 59 respondents at the national and the district levels,
4. Assessment of 4 coastal EIS review meetings at the national level,

5. Assessment of environmental monitoring involving 2 respondents at the district level, and
6. Assessment of perception of technical teams on the use of environmental information in the EIA process involving 4 respondents at the national level.

Assessment of stakeholder participation in the coastal EIA process was conducted through 3 activities:

1. Assessment of perception of technical teams on stakeholder participation in the EIA process involving 4 respondents at the national level,
2. Assessment of stakeholder participation in the 4 coastal EIS review meetings at the national level, and
3. Assessment of stakeholder participation illustrated by a case study involving 89 respondents at the district level.

Research design

- The research was effectively designed and highly representative most of stages of EIA process, as well as encompassing the assessment of the EIA as stated in the legislation and in the practice. It consisted of the assessment of the EIA guidelines and the assessment of the coastal EIA as practiced, which consisted of review of coastal EISs; review of information sources to draft and to review EISs, observation of the coastal EIS review meetings, and review of the environmental monitoring.
- Time utilised for conducting the research and analysing the results amounted to four years.
- The methods used consisted of interviews with open and close-ended questionnaires, observations, site visits, and publication reviews. The checklists included: checklists of parameters to review the EIA guidelines, a checklist of review criteria to assess quality of the coastal EIS, a checklist to observe the coastal EIS review meetings, and checklist of parameters to assess the environmental monitoring.
- The research design was useful because it provided a good insight of the actual practice of the coastal EIA in the Banten Bay area, particularly its strengths and weaknesses. It helps to identify the lapses of the coastal EIA in practice, identify its weaknesses, and its potential effects to the deterioration of the coastal and marine environment.

Case studies

The case studies analysed were carefully selected to be and highly representative of most stages of the coastal EIA process, from the case studies in the Banten Bay area to the national case studies. They are proved helpful to get a more focused analysis of the coastal EIA as practiced in the Banten Bay area and a more general overview of the Indonesian EIA and ICZM situation.

Research findings

- The main findings of this research are as follows:

  - A comprehensive overview of the coastal EIA process as practised in the Banten Bay area, in particular on the use of environmental information and stakeholder participation.
  - The coastal EIA as practised is not effective due to its lack of marine biology information in most of the coastal EIA stages.
  - The coastal EISs reviewed particularly on impact management and monitoring plans have weaknesses and they cannot be used to generate information for ICZM.
  - The coastal EIA process as practised is not likely to generate marine information for coastal monitoring; therefore the provision of marine information should be conducted independently among others through action research.

- Confirmation of the hypotheses:

  - The two hypotheses on "the use of marine environmental information and stakeholder participation in the coastal EIA process has the potential to improve the performance of coastal EIA and ICZM" have been confirmed by the results on lack of marine information and stakeholder participation has lessened the performance of the coastal EIA and thus ICZM.
  - Hypothesis on "the coastal EIA system has the potential effect to be utilised as a management tool to improve ICZM" has been confirmed by the fact that the coastal EIA is the only source of environmental information available to the district government to improve ICZM.

9.5. Suggestions for Further Research

Further research is suggested including:

- Research on the institutional aspects of the coastal EIA and the role of the district governments after the implementation of the regional autonomy law.
- EIA research on the Mining sector in the Banten Bay area since its sea-sand mining activity for the need of materials for the Jakarta Bay Reclamation Project will significantly weaken the environmental quality of the Banten Bay in the coming years.
- EIA research in other similar coastal areas in order to generate a more general overview of coastal EIA and ICZM in the Indonesian situation.

REFERENCES

Abaza, H. (2000). Strengthening future environmental assessment practise: an international perspective. In: *Environmental assessment in developing and transitional countries.* Pp. 271-282. Lee, N. and George, C. (Eds). University of Manchester. John Wiley and Sons Ltd., UK.

Adnan, S., Barrett, A., Nurul Alam, S.M., Brustinow, A. and others (1992). *People's participation, NGOs and the Flood Action Plan: An independent review.* Research and Advisory Services. Report commissioned by Oxfam. Dhaka, Bangladesh.

Ahmad, Y.J. and Sammy, G.K. (1987). *Guidelines to environmental impact assessment in developing countries.* Regional Sea Reports and studies, no 85. UNEP, Nairobi, Kenya. 44 pages.

Archer, J.H. (1989). Resolving intergovernmental conflicts in marine resource management: the US experiences. *Ocean-Shoreline Management.* Volume 12, Pp. 253 -269.

ASEAN/US-COMP/DGF (1993). *The integrated management planning for Segara Anakan, Cilacap, Central Java, Indonesia.* Technical Report no 34. International Centre for Living Aquatic Resources Management (ICLARM), Manila, the Philippines.

Asia Foundation (2002). 1st *Indonesia's Decentralisation Appraisal (IRDA).* The Asia Foundation Report of 2002. Retrieved from: www.asiafoundation.org.

Australian EIA Network (1994). *Review of Commonwealth EIA-Analysis of guideline and procedures in other countries.* http://www.environment.gov.au.portfolio/epg/eianet/eia/os/chapter2.htm

Bankes, N. and Thomson, A.R. (1980). *Monitoring for impact assessment: An analysis of the legal and administrative framework.* Westewater Research Center, the University of British Columbia, UK. 80 pages.

Bappenas / Ford Foundation (1998). *Country Profile Indonesia, 1997: Implementation of Agenda 21.* Bappenas, Jakarta, Indonesia.

Bisset, R. (2000). Methods of consultation and public participation. In: *Environmental assessment in developing and transitional countries.* Pp. 137-147. Lee, N. and George, C. (Eds). University of Manchester. John Wiley and Sons Ltd., UK.

Bisset, R. (1992). Devising an effective environmental impact assessment system for a developing country: the case of the Turks and Caicos Islands. In: *Environmental impact assessment for developing countries*. Biswas, A.K. and Argawala, S.B.C (Eds). Butterworth, UK. 250 pages.

Biswas, A.K. (1992). Summary and recommendations. In: *Environmental impact assessment for developing countries*. Biswas, A.K. and Argawala, S.B.C (Eds). Butterworth, UK. 250 pages.

Biswas, A.K. and Argawala, S.B.C (Eds). (1992). *Environmental impact assessment for developing countries*. Butterworth, UK. 250 pages.

Bodani, L. (2001). *ICZM Stakeholder Participation in Problem Solving*. Retrieved from www.coastalsantren.nu.ac.za/courses/durban/stakeholders.html. Last updated: 12 December 2001.

Bower, B.T. (1994). *A framework for planning for integrated coastal zone management*. NOAA (National Oceanic and Atmospheric Administration), National Ocean Service. Silver Spring, USA. 65 pages.

Borrego. C. (1996). Integrated coastal zone strategy: Need for a more quantitative approach. In: *Coastal Environment: Environmental problems in coastal regions*. Pp. 23-33. Ferrante, A.J. and Brebbia, C.A. (Eds.). Computational Mechanics Publications, Southampton, Boston.

Brown, V.A. (1994). *Acting locally: Meeting the environmental information needs of local government*. Centre for Resource and Environmental studies, Australia National University, Canberra. 67 pages.

Burbridge, P.R., Koesoebiono and Dahuri, R. (1988). Problems and issues in coastal resources management and planning in eastern Sumatra and the Strait of Malacca. *Coastal zone management in the Strait of Malacca*. Pp. 108-117. Burbridge, P.R., Koesoebiono, Dirschal, H. and Patton, B. (Eds). Halifax School of Resource and Environmental Studies, Dalhousie University, Canada.

Caldwell, L.K. (1989). Understanding impact analysis: Technical process, administrative reform, policy principle. In: *Policy through impact assessment: Institutionalised analysis as a policy strategy*. Pp. 7-16. Bartlett, R.V. (Ed). Greenwood Press, New York, USA.

Canter, L.W. and Fairchild, D.M. (1986). Post-EIS environmental monitoring. In: *Methods and experiences in impact assessment*. Pp. 265-285. Becker, H.A. and Porter, A.L. (Eds). D. Reidel Publishing Company, Dordrecht, the Netherlands.

Carpenter, R.A. and Maragos, J.E. (1989). *How to access environmental impacts on tropical islands and coastal areas*. South Pacific Regional Environmental Programme (SPREP) training manual. Environment and Policy Institute, East-West Center, USA. 339 pages.

Cendrero, A. (1989). Land Use Problems, Planning and Management in the coastal zone: An Introduction. *Ocean and Shoreline Management*. Volume 12. Pp. 367-381.

Chua, T.E. (1993). Essential elements of integrated coastal zone management. *Ocean and Coastal Management.* Volume 21. Pp. 81-108.

Christie, P. and White, A.T. (1997). Trends in development of coastal area management in tropical countries: from central to community orientation. *Coastal Management.* Volume 25. Pp. 155-181.

Cicin-Sain, B. (1994). Sustainable development and integrated coastal management. *Ocean and Coastal Management.* Volume 25. Pp. 11-45.

Cicin-Sain, B. and Knecht, R. (1991). Multiple use ocean management in the US: Toward a New Conceptual Framework. *Coastal Zone 91.* Volume 3. American Society of Civil Engineers, New York, USA.

CIDI (Inter American Council for Integrated Development) (2000). *Inter-American strategy for the promotion of public participation in decision making for sustainable development.* Organisation of American States, Unit for Sustainable Development and Environment, USA.

Clark, J.R. (1996). *Coastal zone management: Handbook.* CRC Press, New York, USA. 694 pages.

Crawford, B. (2000). Decentralised coastal and marine resource management in the Philippines: lessons for Indonesia. *Proceedings of the Workshop on Marine and Coastal research in Indonesia.* 13-14 May 2000. Makassar, Indonesia.

Crooks, R and Foley, S. (1995). *Review of environmental assessment and management experience in Indonesia.* Environment management series no 25. World Bank Environment Department, Washington DC., USA.

Dahuri, R. and Dutton, Ian M. (2000). *Integrated Coastal and Marine Management Enters a New Era in Indonesia.* Retrieved from: www.iczm/org/journals/iczm.01/iczmpdfs/iczm_1_4.pdf.

Davies, G.S. (1987). Environmental Impact Assessment, Public Participation and the Project Cycle. *Proceeding of an International Workshop on the Role of Environmental Impact Assessment in the decision-making process.* Pp. 239-252. August 1987. Heidelberg, Federal Republic of Germany.

Doody, J.P., Pamplin C.F., Gilbert C. and Bridge L. (1999). *European Union Demonstration Programme on Integrated Management on Coastal Zones, Thematic Study F- Information required for Integrated Coastal Zone Management.* Report to the European Environment Agency, Copenhagen, Denmark.

Donnely, A., Dalal-Clyton, B. and Hughes, R. (Eds) (1998). *A directory of impact assessment guidelines. 2nd edition.* International Institute for Environment and Development, IIED-WRI-IUCN, U.K.

Doyle, E. (1999). *An evaluation of stakeholder participation in private sector planning for an environmentally sensitive energy development project.* MSc thesis. Institute for Housing and Urban Development Studies, Rotterdam, the Netherlands.

Douven, W.J.A.M., Heun. J.C. and Tiwi, D.A. (2000). Integrated research to support coastal zone management in Banten Bay. *Indonesian Journal of Coastal and Marine Resources.* Volume 3, no 1. Pp. 68-76.

Douven, W.J.A.M. (1999). *Human pressure on marine ecosystems in the Teluk Banten coastal zone: Present situation and future prospects,* Teluk Banten Research Report Series no. 3. IHE, Delft, the Netherlands.

Environment Assessment Branch (1994). *Review of Commonwealth EIA and Public Participation in the EIA Process.* Environment Australia. Australian EIA networks. http://www.environment.gov.au/portfolio/epg/eianet/eia/pubpart/execsum.htm

Environmental Impact Agency (*Bapedal*) (2000). *Decree of Environment of Indonesia no. 8/2000 on public participation and access of information in analysis process of environmental impact assessment. Bapedal,* Jakarta, Indonesia.

FAO (2001). *Technical Consultation on Decentralisation and Rural Development.* A technical report. FAO, Rome, Italy. Retrieved from: www.fao.org/sustdev.

FAO (1991). Development of coastal areas and enclosed seas. *UN Conference on Environment and Development.* Research Paper no 4. UN Food and Agriculture Organisation, Rome, Italy.

Frihy, O.E. (2001). The necessity of environmental impact assessment (EIA) in implementing coastal projects: Lessons learned from the Egyptian Mediterranean Coast. *Ocean and Coastal Management.* Volume 44. Pp. 489-516.

George, C. (2000). Environmental monitoring, management and auditing. In: *Environmental assessment in developing and transitional countries.* Pp. 177-193. Lee, N. and George, C. (Eds). University of Manchester. John Wiley and Sons Ltd., UK.

GESAMP (Group of Experts on the Scientific Aspects Marine Environmental Protection) (1996). *Report of the task force on integrated coastal zone management.* Food and Agriculture Organization of the United nations, Rome, Italy.

Gilpin, A. (1995). *Environmental impact assessment: Cutting edge for the twenty-first century.* Cambridge University Press, UK. 182 pages.

Government of District of Serang (1995a). *Master Plan of Serang District, year 1994: Fact and Analysis.* Government of District of Serang, Indonesia.

Government of District of Serang (1995b). *Master Plan of northern coastal area of Serang District, year 2005.* Government of District of Serang, Indonesia.

Government of District of Serang (1982). *Regional Regulation of Serang District, no 3/1982, on organisation structure and working procedure for Serang District Planning Office.* Government of District of Serang, Indonesia.

Government of Republic Indonesia (1999). *Governmental Regulation no 27/1999 on the Environmental Impact Assessment.* Government of Republic Indonesia, Jakarta, Indonesia.

Government of Republic Indonesia (1993). *Governmental Regulation no 51/1993 on the Environmental Impact Assessment.* Government of Republic Indonesia, Jakarta, Indonesia.

Hale, L.Z. (2000). Achieving integration in coastal management: the challenge of linking national and local levels of government. *Indonesian Journal of Coastal and Marine Resources.* Volume 3, no 1. Pp. 1-8.

Henk, C.G.M. (1987). Incorporation of EIA into existing decision-making procedures in the Netherlands. In: *Proceeding of an International Workshop on the Role of Environmental Impact Assessment in the decision-making process.* Pp. 131-136. August 1987. Heidelberg, Federal Republic of Germany.

Heun, J.C. (1996). *Main conclusions and recommendations (ICZM-research group, phase 1).* IHE, Delft, the Netherlands. Paper presented at workshop on Marine and Coastal Research. November 1996. Jakarta, Indonesia.

Hofwegen, P.J.M. and Jaspers, F.G.W. (1999). *Analytical framework for Integrated Water Resource Management: Guidelines for assessment of institutional frameworks.* IHE Monograph 2. A. A. Balkema, Rotterdam, the Netherlands. 93 pages.

Howlett, D.J.B. and Nagu, J. (1997). *Agriculture Project Planning in Tanzania.* Institute of Development Management, Tanzania and Development Project Planning Centre, University of Bradford, UK.

Hughes, R. (1998). Environmental impact assessment and stakeholder participation. In: *A directory of impact assessment guidelines. 2nd edition.* Pp 21-29. Donnely, A., Dalal-Clyton, B. and Hughes, R. (Eds). International Institute for Environment and Development. IIED-WRI-IUCN, UK.

Ibrahim, A.K.C. (1992). *An Analysis of Quality Control in Malaysian Environmental Impact Assessment Process,* unpublished MSc dissertation. University of Manchester, Manchester, U.K.

IUCN (1997). *Expanding environmental assessment capacity in Sub-saharan Africa: Issues and options.* A discussion paper. Prepared jointly by IUCN and the World Bank. 31 pages.

IUCN (1990). *Caring for the world: A strategy for sustainability.* World Conservation Union (IUCN), Gland, Switzerland. June 1990. 135 pages.

Janicke, M. and Weidner, H. (Eds) (1997). *National environmental policies. A comparative study of capacity building.* UNU World Institute for Development Economic Research (UNU/WIDER), Finland. 320 pages.

Kasim, A. (1993). *Local co-ordination in West Java Province.* Faculty of Administration, University of Indonesia, Jakarta. Indonesia.

Kenchington, R. and Crawford, D. (1993). On the meaning of integration in coastal zone management. *Ocean and Coastal Management.* Volume 21. Pp. 109-127.

Kenchington, R.A. (1990). *Managing Marine Environments.* Taylor and Francis, New York, USA. 248 pages.

Knecht, R.W., Cicin-Sain, B. and Fisk, G.W. (1997). Perceptions of the performance of State coastal zone management program in the United States. *Coastal Management.* Volume 25. Pp. 325-343.

Knecht, R.W. (1994). A commentary on the institutional and political aspects of regional ocean governance. *Ocean and Coastal Management.* Volume 24. Pp. 39-50.

Knecht, R.W. (1993). 'Integration' in the US Coastal Zone Management Program. *Ocean and Coastal Management.* Volume 21. Pp. 183-199.

Kiswara, W. (1999). *Sea grass in Teluk Banten, Indonesia.* Teluk Banten workshop, November 1999. Texel, the Netherlands.

Koesoebiono, Collier, W.L. and Burbridge, P.R. (1982). Indonesia: Resource Use and management in the coastal zone. In: *Man, Land and Sea: Coastal resource use and management in Asia and the Pacific.* Pp. 115-133. The Agricultural Development Council, Bangkok, Thailand.

Lee, N. (2000). Reviewing the quality of environmental impact assessments. In: *Environmental assessment in developing and transitional countries.* Pp. 137-147. Lee, N. and George, C. (Eds). University of Manchester. John Wiley and Sons Ltd., UK.

Lee, N. and George, C. (Eds) (2000). *Environmental assessment in developing and transitional countries.* University of Manchester. John Wiley and Sons Ltd., UK. 290 pages.

Lee, N., Colley, R., Bonde, J. and Simpson, J. (1999). *Reviewing the quality of environmental statements and environmental appraisals.* Occasional paper no 55. EIA Center, Department of Planning and Landscape, University of Manchester, UK. 72 pages.

Lefebvre, C.M. (2000). The people, the coast, and the ocean: Vision 2020, a Coastal Zone 99 retrospective. *Journal for Integrated Coastal zone management* 2000. Pp 25-28. http://www.iczm.org/journals/iczm.01/iczmpdfs.

Maragos, J.E. (1989). Impact of coastal construction on nearshore ecosystem in Oceania: a Review. In: *Coastal Zone Management Handbook*. Clark, J.R. (1996). CRC Press, New York, USA.

Mascia, M.B. (1997). The role of stakeholder participation in the development of Marine Protected Areas in the Caribbean. *Coastal Zone 1997*. American Society of Civil Engineers, New York, USA.

Mera, K. (2000). Decentralisation and Capacity Building: Selecting modes of training for Indonesia. Proceeding in: 47th North American Meetings of the Regional Science Association International. November 9-12, 2000. Chicago, USA.

Ministry of Internal Affairs (1980). *Ministerial Decree of Internal Affairs of Indonesia no 185/1980 on organisation guidance and working procedures for Provincial and District Planning Boards*. Ministry of Internal Affairs, Jakarta, Indonesia.

Ministry of Public Housing, Physical Planning and Environmental Protection (1985). *Handling uncertainty in environmental impact assessment*. Final report. Ministry of Public Housing, Physical Planning and Environmental Protection of the Netherlands.

Ministry of Industry and Trade (1997). *Ministerial decree of Industry and Trade of Indonesia, no 256/1997 on the approval procedure of operational permit for industry, extension permit and industry registration*. Ministry of Industry and Trade, Jakarta, Indonesia.

Ministry of Industry and Trade (1994). *Ministerial decree of Industry and Trade of Indonesia, no 250/1994 on the guidance for environmental impact assessment at Industry sector*. Ministry of Industry and Trade, Jakarta, Indonesia.

Ministry of Communications (1996). *Ministerial Decree of Communications of Indonesia no. 4/1996 on the activities within the Directorate General of Sea Communications required Standard Operation Procedures*. Ministry of Communications, Jakarta, Indonesia.

Ministry of Communications (1996). *Ministerial Decree of Communications of Indonesia no. 4/1996 on the guideline for environmental impact assessment at Sea Communications sector*. Ministry of Communications, Jakarta, Indonesia.

Morvaridi, B. (1998). *Public Consultation – Ineffective link in environmental impact assessment*. Development and Project Planning centre, University of Bradford, UK. 9 pages.

Mostert, E. (1995). *Commissions for environmental impact assessment: Their contribution to the effectiveness of environmental impact assessment*. PhD thesis, Technical University of Delft. Delft, the Netherlands.

Mwalyosi, R., Hughes, R. and Howlett, D.J.B. (1998). *EIA Capacity Development in Africa: An example of a national-led process integrating EIA into project planning in*

Tanzania. Institute for Resource Assessment, Dar-es Salaam and International Institute for Environment and Development, London, UK. 8 pages.

Neame, P.A. (1994). *A Review of the Environmental Impact Assessment process in Indonesia*. EMDI, Dalhouise University, Canada.

Nuraini, S. (1999). *Too little seagrass, too many scoopnetters for groupers and snappers in Teluk Banten, Indonesia*. Teluk Banten workshop. November 1999. Texel, the Netherlands,

ODA (1996). *Manual of Environmental Appraisal*. Overseas Development Administration. Revised July 1996. 131 pages.

OECD (1993). *Coastal Zone Management: Integrated Policies*. Organisation for Economic Co-operation and Development, Paris, France. 125 pages.

Olsen, S. and Hale, L.Z. (1997). Coastal Management: Principles and major features. *Proceedings of the CRMP orientation workshop*. 15-16 January 1997. Puncak, Indonesia. 14 pages.

Organization of American States (OAS) (2001). *Inter-American Strategy for the Promotion of Public Participation in Decision-Making for Sustainable Development (ISP)*. Retrieved from: www.ispnet.org/oas.

Patrick, C. and White, A.T. (1997). Trends in Development of Coastal Management in Tropical Countries: From Central to Community Orientation. *Coastal Management*. Volume 25. Pp. 155-181.

Pernetta, J. and Elder, D. (1993). *Cross-Sectoral, Integrated Coastal Area Planning (CICAP): Guidelines and Principles for Coastal Area Development*. Marine Conservation and Development Report. Gland, Switzerland. 63 pages.

Post, J.C. and Lundin, C.G. (Eds) (1996). *Guidelines for Integrated Coastal Zone Management*. Sustainable Development Studies and Monographs Series no 9. The World Bank, Washington, DC., USA. 15 pages.

Procoast (2000). *Proceedings of the interregional seminar on the harmonisation of uses and interests in the Baltic Sea Coastal Zone*. 11-13 September 2000. Eckernforde, Germany. 101 pages.

Rijsberman, F.R. (Ed) (1999). *Conflict management and consensus building for integrated coastal management in Latin America and the Caribbean*. Environment Division of the Sustainable Development Department of the Inter-American Development Bank. http://www.iadb.org/publications. 60 pages.

Rijsberman, F.R. (1993). The effectiveness of the Helsinki Convention as a tool for the integrated coastal resources management of the Baltic Sea. In: *OECD, Coastal Zone Management: Selected case studies*. Pp. 283-310. OECD, Paris, France.

Sadler, B. (1996). *Environmental Assessment in a Changing World: Evaluating Practice to Improve Performance.* International Study on the Effectiveness of Environmental Assessment, Final Report.

Sadler, B. and Verheem, R. (1996). *The Status of EIA systems: Results of an International survey.* Country status reports on Environmental Impact Assessment. The Netherlands Commission for Environmental Impact Assessment, the Netherlands.

Scholten, J.J. (1995). Reviewing EISs/EIA reports. *Report of the EIA process strengthening workshop.* International study of the effectiveness of environmental assessment. 4-7 April 1995. Canberra, Australia.

Scholten, J.J. (1992). Contribution of EIA to decision-making: experiences from the Netherlands. In: *Environmental Impact Assessment for developing countries.* Pp. 163-167. Biswas, A. K, Agarwala, S.B.C. (Eds). Butterworth, UK.

Sloan, N.A. and Sugandhy, A. (1994). An Overview of Indonesian Coastal Environmental Management. *Coastal Management.* Volume 22. Pp. 215-233.

Smith, L.G. (1993). *Impact Assessment and Sustainable Resource Management.* Longman Group, UK. 210 pages.

Soet, M.C. (1987). First experiences with public participation in EIA. *Proceeding of an International Workshop on the Role of Environmental Impact Assessment in the decision-making process.* Pp. 253-260. August 1987. Heidelberg, Federal Republic of Germany.

Sorensen, J.C. (1997). National and international efforts at integrated coastal zone management definitions, achievement, and lessons. *Coastal Management.* Volume 25. Pp. 3-41.

Sorensen, J.C. (1993). The international proliferation of integrated coastal zone management efforts: Definitions, achievements, and lessons. *Ocean and Coastal Management.* Volume 21. Pp. 3-41.

Sorensen, J.C. and Mc. Creary, S.T. (1990). *Coasts: Institutional arrangement for managing coastal resources and environment.* Coastal Management Publication No 1, USAID series. National Park Service, Office of International Affairs, Washington, DC., USA. 194 pages.

Spooner, B. (1998). Review of the quality of the EIA guidelines, their use and circumnavigation. In: *A directory of impact assessment guidelines. 2nd edition.* Pp. 15-20. Donnely, A., Dalal-Clyton, B. and Hughes, R. (Eds). International Institute for Environment and Development, IIED-WRI-IUCN, UK.

State Ministry of Environment (2000a). *Ministerial Decree of Environment of Indonesia no. 41/2000 on the guidance for the formation of evaluation committee for Environmental Impact Statement at district and city level.* State Ministry of Environment, Jakarta, Indonesia.

State Ministry of Environment (2000b). *Ministerial Decree of Environment of Indonesia no. 40/2000 on the working procedure for evaluation committee for Environmental Impact Statement.* State Ministry of Environment, Jakarta, Indonesia.

State Ministry of Environment (2000c). *Ministerial Decree of Environment of Indonesia no. 2/2000 on the guidance for evaluation of Environmental Impact Statement.* State Ministry of Environment, Jakarta, Indonesia.

State Ministry of Environment (1999). *Ministerial Decree for the Environment no. KEP-30/MENLH/10/1999 on the Indonesian EIA general guideline.* State Ministry of Environment, Jakarta, Indonesia.

State Ministry of Environment (1996a). *Agenda 21-National Strategy for Sustainable Development.* State Ministry of Environment, Jakarta, Indonesia.

State Ministry of Environment (1996b). *Indonesia's Marine Environment, A Summary of Policies, Strategies, Actions and Issue.* State Ministry of Environment, Jakarta, Indonesia.

Stephenson, D., Brooke, C. and Nixon, A. (Eds.) (1994). *Public participation in EIA: a Review of Experience in Europe and the UK.* Center for Environmental Management and Planning (CEMP), Aberdeen, UK. 76 pages

Taylor, M.C. (1993). *Integrated Environmental Development Program (INREDEP).* Final Report. The Ministry of Environment of Indonesia and Dalhouise University, Canada.

Tiwi, D.A., Douven, W.J.A.M. and Rijsberman, F.R. (2001a). The changes of powers, its influence to ICZM in Banten Bay, Indonesia. *Presented in the Conference on People and the Sea Maritime Research (MARE).* 31 August-1 September 2001. Amsterdam, the Netherlands.

Tiwi, D.A., Douven, W.J.A.M. and Rijsberman, F.R. (2001b). Decentralisation of power, implementation of the EIA and the ICZM in Banten Bay, Indonesia. *Poster presentation in the Conference on Coastal Zone 2001.* 16-18 July 2001.Cleveland, USA.

Tiwi, D.A., Douven, W.J.A.M. and Rijsberman, F.R. (2001c). The practice of the EIA in the Banten Bay, Indonesia. *Poster presentation on the Conference on Challenges of a Changing Earth.* July 10-13, 2001. Amsterdam, the Netherlands. Received award on best poster presentation, category Coasts and Oceans.

Tiwi, D.A, Douven, W.J.A.M. and Rijsberman, F.R. (2001d). The EIA in Banten Bay, Indonesia. *Proceedings of the Second interregional seminar on the harmonisation of uses and interests in the Baltic Sea Coastal Zone.* 23-25 April 2001. Fredericia, Denmark.

Tiwi, D.A. (2000). ICZM in Banten Bay, Indonesia. *Proceedings of the workshop on Prospect of integrated marine tourism development, corridor: Jakarta-Banten-Lampung.* 18-19 October 2000. Jakarta, Indonesia.

Tiwi, D.A., Douven, W.J.A.M., and Rijsberman, F.R. (2000). MIS to support ICZM in Banten Bay, Indonesia. *Proceedings of the workshop on Marine and coastal research in Indonesia.* 13-14 May 2000. Makassar, Indonesia.

Tiwi, D.A. (1999). Descriptions and analysis of institutional problems in the Banten Bay coastal zone. *Proceeding of the 4th Indonesian Students' Scientific Meeting on Energy and Environment.* 8-9 October 1999. Kassel, Germany.

Tiwi, D.A. (1998). Problems in Banten Bay coastal zone management. *Proceedings of the Workshop on Small Islands Development.* 2-3 December 1998. Jakarta, Indonesia.

Turner, R.K. and Adger, N. (1996). *Coastal zone resources assessment guidelines.* Land-Ocean Interaction in the Coastal Zone (LOICZ) reports and studies no 5. Texel, the Netherlands.

Uhel, R. (1999). *Knowledge and Information as the basis for ICZM: needs and instruments – role of the EU institutions.* Paper based on the Theme F study of the Demonstration programme – Information required for ICZM. Brussel, Belgium.

UNEP (1995). *Guidelines for integrated management of coastal and marine areas.* Regional Sea Reports and studies, no 161. UNEP, Nairobi, Kenya.

UNEP (1990). *An approach to environmental impact assessment for projects affecting the coastal and marine environment.* Regional Sea Reports and studies, no 122. UNEP, Nairobi, Kenya. 35 pages.

UNEP (1988). *Environmental in assessment: Basic procedure for developing countries.* Regional office for Asia and the Pacific. UNEP, Nairobi, Kenya. 13 pages.

University of Indonesia (1995). *Research study on the collection and interpretation system of EIA data.* Final reports. PPSML-LPUI, Jakarta, Indonesia.

Vallejo, S.M. (1994). New structures for decision-making in integrated ocean policy. In: *Ocean Governance: Sustainable development of the sea.* Pp. 71-96. Payoyo, P. B. (Ed). The United Nations University.

Vallejo, S.M. (1993). The integration of coastal zone management into National Development Planning. *Ocean and Coastal Management.* Volume 21. Pp. 163-182.

WCC (1993). *Preparing to meet the coastal challenges of the 21st century.* World Coast Conference Report. World Coast Conference 1993. Noordwijk, the Netherlands.

Werner, G. (1992). Environmental impact assessment in Asia: Lessons from the past decade. : *Environmental impact assessment for developing countries.* Biswas, A.K. and Argawala, S.B.C (Eds). Butterworth, UK. 250 pages.

White, A.T. (1989). Comparison of Coastal resources Planning and Management in the Asean countries. 6th Symposium on Coastal and Ocean Management (Charleston,

South Caroline, 1989). *Coastal Zone 89.* Pp. 2123-2133. American Society of Civil Engineers, New York, USA.

Wood, C. (1995). *EIA: a comparison review.* The EIA centre, Department of Planning and Landscape, University of Manchester, UK. 337 pages

World Bank (1999). Public consultation in the EIA process: A strategic approach. *Environmental Assessment Source book update*, no 26. May 1999. World Bank, Washington, D.C., USA.

World Bank (1994). Coastal zone management and environmental assessment. *Environmental assessment Source book update*, no 7. March 1994. World Bank, Washington, DC., USA.

World Bank (1993). *Indonesia Environment and Development: Challenges for the Future*, Environment Unit, Country Department III, East Asia and Pasific Region.

World Bank (1993). The World Bank and environmental assessment: An overview. *Environmental assessment Source book update*, no 1. April 1993. World Bank, Washington, DC, USA.

World Summit (2002). The World Summit on Sustainable Development. *Plan of Implementation*. Johannesburg, South Africa.

Zieren, M., Kinaly, K., Darussalam W. and Ramli, A. (1998). Coastal zone management issues in Riau, and the need for large-scale solutions and centralised actions. *Proceedings of the national conference on the Management of Indonesian Marine and Coastal Resources*. March 1998. Bogor, Indonesia.

Zulhasni (2000). Country Studies of EA: EIA in Indonesia. In: *Environmental assessment in developing and transitional countries.* Pp. 205-210. Lee, N. and George, C. (Eds) University of Manchester. John Wiley and Sons Ltd., UK.

SAMENVATTING

INTEGRAAL KUSTBEHEER EN MILIEU-EFFECT RAPPORTAGES

Integraal kustbeheer (Integral Coastal Zone Management) wordt reeds jaren gepropageerd als een manier om conflicten tussen gebruikers van mariene hulpbronnen te verminderen en om nadelige effecten van menselijke activiteiten in kustzones op kust en mariene ecosystemen te verminderen. Integraal kustbeheer is een proces dat uitgaat van de interactie tussen het natuurlijke en menselijke systeem, de balans tussen economie, ontwikkeling en milieu, de bescherming van sociale structuren, en de interactie en consensus ontwikkeling tussen actoren. Kustbeheer programma's kunnen door het gebruik van Milieu Effect Rapportages (MER) (Environmental Impact Assessment) worden versimpeld en zodoende meer effectief worden. De MER procedure voor economische ontwikkelingen in kustgebieden kan zodoende worden gebruikt als een effectief instrument om het kustbeheer te ondersteunen met name bij het bepalen van effecten van kustontwikkelingen die potentieel nadelig zijn voor kust en marine omgevingen.

Om daadwerkelijk integraal kustbeheer door middel van MER's te kunnen verbeteren moet aan een aantal voorwaarden worden voldaan, onder meer: een goede wettelijke basis, voldoende mariene informatie, adequate betrokkenheid en deelname van actoren en voldoende institutionele capaciteit. Met name voorziening van milieu informatie en betrokkenheid en deelname van actoren zijn twee belangrijke elementen binnen de MER die een belangrijke rol spelen bij het verbeteren van integraal kustbeheer. Wanneer beiden op de juiste wijze worden uitgevoerd biedt dit belangrijke voordelen voor het verbeteren van de effectiviteit van de MER. Dit geldt met name voor vier belangrijke fases binnen de MER procedure: scoping, evaluatie van significantie van effecten, evalueren van MER rapporten en het implementeren van het MER besluit.

INTEGRAAL KUSTBEHEER EN MER IN INDONESIE

Integraal kustbeheer is veelvuldig genoemd als een benadering om de druk van economische ontwikkelingen in kustgebieden in landen zoals Indonesië te verminderen. Integraal kustbeheer in Indonesië is echter niet effectief geïmplementeerd. Hiervoor zijn in de literatuur verschillende redenen aangegeven: gebrek aan communicatie en coördinatie tussen overheidsinstanties, conflicten binnen de wetgeving, een niet effectieve implementatie van wetgeving en een te beperkte integratie en coördinatie binnen en tussen sectoren. Een effectief programma voor vergunningverlening van economische ontwikkelingen in kustgebieden door middel van het gebruik van MER zou integraal kustbeheer effectiever maken.

Het koppelen van de MER aan het systeem van vergunningverlening wordt als vitaal gezien voor de implementatie van milieustandaarden en -regulering in Indonesië. Er zijn echter een aantal obstakels die aanzienlijke tijd en energie zullen kosten om te overkomen. Om de MER meer effectief en bruikbaar te maken voor beleidsmakers zouden de procedures transparanter moeten zijn om zodoende implementatie van beslissingen te vergemakkelijken. Ook in Indonesië worden de voorziening van milieu-informatie en de betrokkenheid en deelname van actoren beschouwd als belangrijke elementen binnen het integratie proces van MER's in kustgebieden en integraal kustbeheer. Een voor-analyse uitgevoerd binnen deze studie in het kustgebied van de baai van Banten (West-Java) laat zien dat dit een belangrijke rol kan vervullen bij het verbeteren van integraal kustbeheer. Het adresseren van huidige en toekomstige nadelige effecten van ontwikkelingen rond Banten Baai vraagt om actievere interventies van de district overheid, met name het District Planning Bureau (Bappeda). Een meer effectief gebruik van de MER kan dit stimuleren. De recente politieke en institutionele veranderingen in Indonesië bieden hiervoor mogelijkheden.

BIJDRAGE VAN MILIEU-INFORMATIE VOORZIENING EN DEELNAME VAN ACTOREN AAN DE MER

Milieu-informatie is essentieel om negatieve invloeden van kustontwikkelingen vast te stellen en te evalueren. Goede milieu-informatie vergroot ook het besef en inzicht van beleidsmakers in potentiële milieu problemen in kustgebieden. Daarnaast, levert milieu-informatie belangrijke input voor het meten en de evaluatie van effecten van uitgevoerde projecten, aan de hand waarvan toekomstige plannen aangepast kunnen worden. Om effectiever te zijn zou de MER in kustgebieden echter meer kust en mariene informatie moeten bevatten.

Betrokkenheid en deelname van actoren zal het vertrouwen onder de bevolking vergroten dat negatieve effecten van kustontwikkelingen zijn meegenomen in studies en afwegingen. Voor een effectieve en efficiënte MER procedure zullen de overheid en gemeenschap moeten samenwerken. Dit zal leiden tot een beter beleidsvormingsproces. Alle actoren die geïnteresseerd zijn bij een ontwikkelingsproject moeten betrokken worden bij de MER procedure, moeten regelmatig deelnemen, en moeten ook toegang hebben tot de relevante informatie.

PROBLEEMSTELLING EN HYPOTHESE

De bovenstaande analyse leidt tot de volgende centrale probleemstelling van de studie:

- Hoe kan milieu-informatie voorziening en deelname van actoren bijdragen tot een meer effectief gebruik van MER's in kustgebieden en integraal kustbeheer?
- Hoe kan een betere milieu-informatie voorziening en deelname van actoren georganiseerd worden door de district overheid?

De hypothese van deze studie luidt:

- Het gebruik van mariene milieu-informatie in MER procedures in kustgebieden heeft de potentie om de uitvoering van MER's in kustgebieden en integraal kustbeheer te verbeteren.
- Betrokkenheid en deelname van actoren in MER procedures in kustgebieden heeft de potentie om de uitvoering van MER's in kustgebieden en integraal kustbeheer te verbeteren.
- Het MER systeem als management instrument heeft de potentie om integraal kustbeheer te verbeteren.

- Milieu-informatie gegenereerd door een multidisciplinair onderzoeksteam draagt bij aan betere ondersteuning van managementbeslissingen in het kader van integraal kustbeheer.
- De politieke veranderingen binnen de Indonesische overheid hebben de potentie de rol van de district overheid binnen MER's in kustgebieden en integraal kustbeheer te verbeteren.

METHODEN VAN ONDERZOEK

In het onderzoek zijn de MER richtlijnen zoals neergelegd in een aantal (internationale en nationale) documenten vergeleken met de praktijk. Het onderzoek beperkte zich tot die overheidssectoren die het meest van belang zijn voor kustontwikkelingen in Indonesië, namelijk de zee communicatie sector en de industrie sector. Het analyseren van het gebruik van milieu-informatie binnen de MER procedure bestond uit 6 onderzoeksactiviteiten:

1. Analyse van milieu informatie in 3 MER richtlijnen,
2. Analyse van milieu informatie beschreven in 10 kust MER's,
3. Analyse van milieu informatie bronnen om de MER op te stellen en te evalueren,
4. Analyse van 4 kust MER bijeenkomsten op nationaal niveau,
5. Analyse van milieu metingen op district niveau,
6. Analyse van percepties van technische teams met betrekking tot het gebruik van milieu informatie binnen de MER procedure.

De analyse van de betrokkenheid en deelname van actoren in de MER procedures besloeg 3 onderzoeksactiviteiten:

1. Analyse van de percepties van technische teams met betrekking tot deelname van actoren,
2. Analyse van actor deelname in 4 kust MER bijeenkomsten op nationaal niveau,
3. Analyse van actor deelname in een casestudy op district niveau.

Data is verzameld door middel van interviews, vragenlijsten, observaties, veld bezoek, en analyse van rapporten. Checklists zijn gebruikt om parameters in MER richtlijnen te analyseren, de kwaliteit van MER's te beoordelen, MER bijeenkomsten te observeren, en parameters van milieu metingen te analyseren.

Deze studie is uitgevoerd in het kader van het KNAW/NWO/WOTRO Teluk Banten Onderzoeksprogramma, een Indonesisch-Nederlands onderzoeksprogramma op het gebied van integraal kustbeheer in Banten Baai, Indonesië (1997-2001).

HET GEBRUIK VAN MILIEU-INFORMATIE EN ACTOR DEELNAME IN DE KUST MER PROCEDURES IN BANTEN BAAI

- ***Beperkte beschikbaarheid van milieu informatie***
  In de meeste fasen van de onderzochte MER procedures was het gebruik van milieu informatie beperkt. De beschikbaarheid van belangrijke informatie voor MER procedures zoals milieu data, monitoring gegevens en de status van natuurlijke hulpbronnen was beperkt. De onderzochte MER richtlijnen boden slechts beperkte indicatoren in de groep van de mariene biologie. Activiteiten zoals het opstellen en evalueren van rapporten en het meten van effecten na

ontwikkeling gingen niet genoeg in op milieu data. De kwaliteit van de geëvalueerde kust MER's was niet voldoende om te komen tot een goed afgewogen goedkeuring. Basis informatie was niet beschikbaar en wanneer beschikbaar vaak moeilijk te gebruiken binnen de MER procedure.

- ***Beperkte uitwisseling van data***
  Uitwisseling van data is belangrijk voor de compilatie en evaluatie van een MER en een kustbeheersplan. Deze uitwisseling vond maar in beperkte mate plaats tussen overheidskantoren op lokaal, provinciaal en nationaal niveau. De district overheid had maar beperkte toegang tot lokale informatie over mariene hulpbronnen die vaak verzameld was door de provinciale en nationale overheden of onderzoeksinstellingen. De data standaarden voldeden vaak niet aan de behoeftes van de district overheid. Daarnaast was er gebrek aan een methodologie om informatie uit te wisselen. Eveneens werd ook gebrek aan communicatie tussen consultants betrokken bij de MER studies gesignaleerd.

- ***Tekortkomingen van de MER in kustgebieden***
  De mogelijkheden om de kust MER te reviseren waren beperkt omdat de evaluatie van de MER door de district overheid plaatvond na indiening bij de centrale MER commissie. De gespecificeerde evaluatie criteria verminderden de mogelijkheid tot verbeteren van de kwaliteit van de kust MER. De impact evaluatie methoden gebruikt in de MER waren in sommige gevallen beperkt en vaak meer kwalitatief dan kwantitatief van aard. In een aantal gevallen werd de data beschreven in de MER's niet geanalyseerd. De evaluatie van negatieve effecten van de voorgestelde ontwikkeling was in enkele kust MER's niet juist geanalyseerd.

- ***Niet voldoende informatie management***
  Informatie beheer was beperkt vanwege:
  - De beperkte beschikbaarheid van lokale milieu gegevens, milieu monitoring gegevens en de status van kust hulpbronnen.
  - Gebrekkige uitwisseling van informatie omdat de informatie stroom tussen en binnen kantoren beperkt was.
  - Gebrek aan beleid en gebrek aan (financiële) middelen binnen de district overheid onder meer in het geval van kust monitoring.
  - Fragmentatie van verantwoordelijkheid en het bestaan van conflicten in kust en mariene beheer in kustgebieden met betrekking tot vergunningen en MER goedkeuringen tussen nationale en locale instituten.
  - Onvoldoende onderzoek in het studiegebied naar de stand van zaken van het kustmilieu en effecten van kust ontwikkelingen hierop.

- ***Het gebrek aan macht van de deelnemende actor***
  In de meeste onderzochte MER's was de deelname van actoren beperkt. De actoren waren alleen betrokken bij het aanleveren van data en informatie gedurende de scoping fase, terwijl ze beperkte toegang hadden tot informatie over het ontwikkelingsplan en de MER procedure. De deelname van actoren in het evalueren van de MER was eveneens zeer beperkt en leidde in geen enkel onderzocht geval tot een andere eindbeslissing. Bekendheid van de actoren met het project, communicatie met de ontwikkelaar en deelname in de

goedkeuring was laag. Kennis van actoren over de mogelijke milieu-effecten van kustprojecten werd niet meegenomen in de MER procedure.

- ***Beperkingen evaren door actoren***
  De resultaten en bevindingen van de onderzochte kust MER's waren ontoegankelijk voor de actoren. De gehanteerde taal tijdens MER evaluatie bijeenkomsten - met veel technische termen - was een beperking voor de actoren om de MER te begrijpen. Voor projecten die waren goedgekeurd op nationaal niveau of die waren gefinancierd op dat niveau, werd meestal een nationale consultant gecontracteerd. De geografische afstand tussen de consultant en de lokale actoren beperkte hun contact en communicatie.

CONCLUSIES

Het onderzoek in Banten Baai wijst uit dat verbeteringen van het gebruik van milieu-informatie en van de deelname van actoren in de MER procedures in kustgebieden essentieel zijn om te komen tot een beter integraal kustbeheer. In het onderzoek worden een aantal suggesties voor deze verbeteringen gegeven.

Hieronder volgt een reflectie op de centrale probleemstellingen.

Het gebruik van mariene milieu-informatie en betrokkenheid en deelname van actoren in de MER procedure in kustgebieden heeft de potentie om de uitvoering van MER's in kustgebieden en integraal kustbeheer te verbeteren
De MER procedure in kustgebieden kan gebruikt worden als een beheersinstrument om integraal kustbeheer te verbeteren. Door een betere wetgeving en technische richtlijnen, betere institutionele capaciteit, betere deelname van actoren, betere uitwisseling en disseminatie van kust en mariene informatie en betere monitoring krijgt de districtsoverheid meer ondersteuning bij de analyse en goedkeuring van MER's in kustgebieden. Dit zal de kwaliteit van MER's in kustgebieden significant verbeteren.

Milieu-informatie gegenereerd door een multi-disciplinair onderzoeksteam draagt bij aan betere managementbeslissingen in het kader van integraal kustbeheer.
Studie naar de natuurlijke en menselijke dimensie van het kustsysteem is nodig, inclusief studie naar hoe de hieruit resulterende informatie bruikbaar te maken voor kustbeheerders. Banten Bay Informatie Systeem (BBIS) is een prototype informatie systeem dat informatie geeft over het kustsysteem van Banten Baai. BBIS assisteert de district overheid in Banten bij het uitvoeren van MER's en van het integrale kustbeheer. Het systeem is in nauwe samenwerking met de district overheid ontwikkelt en bevat informatie uit het Teluk Banten Onderzoeksprogramma. BBIS biedt de overheid geïntegreerde milieu-informatie met betrekking tot geselecteerde kust en mariene aspecten (zeegras, vis, koraalriffen en vogels) die kunnen dienen als parameters voor milieu monitoring. Door deze gegevens op gestructureerde wijze beschikbaar te maken draagt BBIS bij aan een beter mariene en kust beheer in Banten.

De politieke veranderingen binnen de Indonesische overheid hebben de potentie de rol van de district overheid binnen MER's in kustgebieden en integraal kustbeheer te verbeteren
De recente veranderingen in het Indonesisch beleid, met name de decentralisatie in 1999, heeft de implementatie van wetten en beleid zoals de MER regulatie sterk

beïnvloedt. Zo zijn er reglementen aangenomen over de MER evaluatie procedures, over de opstelling van regionale MER commissies en over de deelname van actoren in de te volgen procedures. Deze reglementen hebben de rol van de district overheid in de goedkeuring van MER's sterk verbeterd. Ook de nieuwe Provincie Banten, waar de baai van Banten binnen valt, heeft de macht van de district overheid met betrekking tot het beheer van mariene en kust gebied sterk verbeterd. Deze verbeterde rol voor de district overheid in integraal kustbeheer verhoogt de behoefte aan effectieve MER's op district niveau. De resultaten van de onderliggende studie geven aan dat door verbeterde milieu-informatie en het verbeteren van de deelname van betrokkenen in MER procedures in kustgebieden, de district overheid een beter ondersteuning krijgt om dit instrument te gebruiken ten behoeve van integraal kustbeheer. Locaal zelfbestuur, een actieve deelname van de district overheid, en integraal kustbeheer zijn verder gestimuleerd door het instellen van een nationaal ministerie van Marine Zaken en Visserij.

ACKNOWLEDGEMENTS

This thesis would not have been possible without wide ranging support from many parts.

The Foundation for the Advancement of Tropical Research (WOTRO) and the Royal Netherlands Academy of Arts and Sciences (KNAW) of the Netherlands provided the core funding. Ir. Jetze C. Heun of UNESCO-IHE, the project manager of the Teluk Banten Research Programme, provided the moral and practical support for my research.

My promoters, Prof.Dr. Wim van Viersen and Prof.Dr.Ir. Frank R. Rijsberman for their encouragement and spirits in guiding my research. My co-promotor, Dr. Wim Douven for his valuable support during the study. The members of the review committee, Dr. Ir. Indroyono Soesilo (University of Indonesia), Prof.Dr. Peter Burbridge (University of Newcastle), Prof.Dr. Klaas-Jan Beek (ITC Enschede) and Prof.Dr. Arthur Mol (Wageningen University), for their precious comments in improving my thesis.

The UNESCO-IHE Institute for Water Education in Delft, the Netherlands, for providing me with a home base, support and access to valuable resources. The International Water Management Institute (IWMI) in Colombo, Sri Lanka, for providing support during my visits to my promoter.

My office, the Agency for the Assessment and Application of Technology (BPPT), in particular the Center of Technology Assessment for Natural Resources Management in Land and Coastal-Small Islands (Jakarta, Indonesia) for giving me the opportunity to undertake this study.

Most important of all, government staff of Serang District and of the Sea Communications and Industry sectors and many others who generously provided me with valuable information. My research collegues, including Siti Nuraini, Wawan Kiswara, Yus Rusila Noor, Nyoman Sukmantalya, Ton Hoitink. My editors, Mr. Nilman Fernando and Ms. Mayanthie from IWMI-Srilanka for their help in editing the manuscript of this thesis. My family, Aunur Rofiq Hadi, Audina Furi Nirukti and Candrika Dini Khairani for their love, patience and support. My warmest thanks and regards go to them all.

CURRICULUM VITAE

Dwi Abad Tiwi was born on 7 October 1956 in Yogyakarta, Indonesia. She spent her education in her hometown and studied her undergraduate from Gajah Mada University, in the field of Architectural Engineering from 1975 until 1982. Her Master in Urban Planning (MUP) dealt with urban and regional planning and was earned from Michigan State University, USA in 1989 after 2 years study. From 1983 until now she works at the Agency for the Assessment and Application of Technology (*Badan Pengkajian dan Penerapan Teknology, BPPT*) in Jakarta, Indonesia. She has been involved in various researches dealing with environment, urban and regional planning, coastal and small islands. She co-ordinated Indonesian research activities of Teluk Banten Research Programme, a research collaboration between Indonesia and the Netherlands from 1997-2001, in which she used her research on EIA and ICZM in Banten Bay as her thesis. She published several papers and was involved in various national and international seminars. Her research received an international award on Best Poster Presentation, category Coasts and Oceans (IGBP Conference, 2001).

APPENDIX I

Proposed Addition of Indicators for EIA Technical Guideline of the Sea Communications sector

(based on Ministerial decree: *Kepmen Perhubungan* no. KM5 Tahun 1966)

1. Addition of parameters and indicators

1.1. Baseline conditions

3. **Description of environmental Component**
3.1. Land
3.1.1. Land Status: Ownership: state or community
3.1.2. Land use: - Land use planning - Existing land use - Land and natural resources inventory - Land cover - Land utilisation based on Spatial Plan
3.1.3. Hazard on natural disaster: Sensitivity of the disaster: - erosion - flooding - earthquake
3.1.4. Physic, chemical characteristics and type of soil: - Physical characteristics (texture, structure, sensitivity to erosion, etc) - Chemical characteristics (pH, ferrous metal. etc) - Soil type
3.2. Water Resource (surface and groundwater)
3.2.1 Water resources and water quality
3.2.2. Average debit
3.2.3. Water allocation
3.2.4. Distance with location of activities
3.2.5. Physical and chemical characteristics of water
3.3 Hydro-oceanography
3.3.1 Wave height
3.3.2 Tidal pattern
3.3.3 Current pattern
3.3.4 Bathymetry
3.4. Air and Noise
3.4.1. Air Quality: - Gas and dust - Noise and noise resources
3.4.2. Climate condition: - Temperature - Humidity - Rain fall - Wind direction and wind speed
3.5. Flora and fauna
3.5.1. Land flora and fauna: - Number of types of flora and fauna - Number of types of rare species of flora and fauna
3.5.2. Terrestrial flora and fauna:- Number of types of flora and fauna - Number of types of rare species of flora and fauna
3.6. Social Economic and Culture
3.6.1. Population :- Total population - Population density - Population growth - Population structure

- Number of population affected by the plan - Infrastructure affected by the plan
3.6.2. Socio-economy: - Average income - Jobs - Employment opportunity
3.6.3.Socio-culture: - Culture value - Culture conflict - Culture diversity - Public infrastructure affected by the plan - Perception to the plan
3.7. Others
3.7.1. Sensitive/Critical Area:- Conservation forest - Nature Conservation area - Culture Conservation area - Tourism area
3.7.2 Spatial Plan: - Location and extent of the plan
3.7.3 Other activities within location of the Plan: - Harbours - Housing/settlements - Industries - Public utilities - Linkages with other activities

1.2. Environmental impact

3. Description of the affected environment
3.1. Land
3.1.1. Land Status: Ownership: state or community
3.1.2. Land use: - Land use planning - Existing land use - Land and natural resources - Land cover - Land utilisation based on Spatial Plan
3.1.3. Hazard on natural disaster: Sensitivity of the disaster: - erosion - flooding - earthquake
3.1.4.Physic, chemical characteristics and type of soil: - Physical characteristics (texture, structure, sensitivity to erosion, etc) - Chemical characteristics (pH, ferrous metal. etc) - Soil type
3.2. Water Resource (surface and groundwater)
3.2.1 Water resources and water quality
3.2.2. Average debit
3.2.3. Water allocation
3.2.4. Distance with location of activities
3.2.5. Physic and Chemical characteristics of water
3.3 Hydro-oceanography
3.3.1 Wave height
3.3.2 Tidal pattern
3.3.3 Current pattern
3.3.4 Bathymetry
3.4. Air and Noise
3.4.1. Air Quality: - Gas and dust - Noise and noise resources
3.4.2. Climate condition:- Temperature - Humidity - Rain fall

- Wind direction and wind speed
3.5. Flora and fauna
3.5.1. Land flora and fauna: - Number of types of flora and fauna - Number of types of rare species of flora and fauna
3.5.2. Terrestrial flora and fauna:- Number of types of flora and fauna - Number of types of rare species of flora and fauna
3.6. Social Economic and Culture
3.6.1. Population : -Total population - Population density - Population growth - Population structure - Number of population affected by the plan - Infrastructure affected by the Plan
3.6.2. Socio-economy: - Average income - Jobs - Employment opportunity
3.6.3. Socio-culture: - Culture value - Culture conflict - Culture diversity - Public infrastructure affected by the plan - Perception of community on the plan
3.7. Others
3.7.1. Sensitive/Critical Area:- Conservation forest - Nature Conservation area - Culture Conservation area - Tourism area
3.7.2 Spatial Plan: - Location and extent of the Plan
3.7.3 Other activities within location of the Plan: - Harbours - Housing/settlements - Industries - Public utilities - Linkages with other activities

2. Institutions for impact mitigation measures and impact monitoring

Activities	Proposed Institution
Executing institution for impact mitigation	Initiator
Monitoring institution for impact mitigation	- EIA provincial/district offices - Provincial and District Planning offices - Provincial Sea Communications office - Environmental research centre - NGO
Related institution for impact mitigation	- Health District office - District Sea Communications office
Executing institution for environmental monitoring	Initiator
Monitoring institution for environmental impact monitoring	- EIA provincial/district offices - Provincial and District Planning offices - Provincial Sea Communications office - Environmental research centre - NGO
Related Institution for environmental impact monitoring	- Health District office - District Sea Communications office

APPENDIX II

Proposed Addition of Indicators for EIA Technical Guideline of the Industry sector

(based on Ministerial decree: *Kepmen Indag* no. 250 / 1994)

1. Addition of parameters and indicators

1.1. Baseline condition

3. Description of Environmental Component
3.1. Spatial Component
3.1.1. Land use: - Master Plan - Detail Plan
3.1.2. Transportation System: - Public transportation lines - Traffic density
3.1.3. Settlement and Development Planning: - Area and location of settlement - Number of housing
3.1.4. Public Facilities:- Public schools - Health facilities - Market - Religion facilities
3.1.5 Type and location of activities surrounding
3.2. Physical-Chemical Environmental Component
3.2.1. Climate: - Temperature - Humidity
3.2.2Air Quality: - Wind direction and speed - Gases (CO, NOx, etc) - Dust - Noise
3.2.3. Physiography: - Topography - Geology (morphology, aquifer)
3.2.4. Hydrology
3.2.5. Hydro-oceanography
3.3. Biological Component
3.3.1. Land : - Type and bio-diversity of flora - Type and bio-diversity of fauna
3.3.2. Marine: - Type and bio-diversity of flora - Type and bio-diversity of fauna
3.4. Socio-economic and culture components
3.4.1. Demography: - Total population - Population density - Population growth and structure - Total population affected by the plan
3.4.2. Socio Economic: - Average income - Jobs - Employment opportunity - Regional income
3.4.3. Socio Culture: - Culture value - Culture conflict - Culture diversity - Public facilities affected by the plan - Perception of community on the plan

1.2. Environmental impact

3. Description of the affected environment
3.1. Spatial Component
3.1.1. Land use: - Master Plan - Detail Plan
3.1.2. Transportation System: - Public transportation lines - Traffic density
3.1.3. Settlement and Development Planning: - Area and location of settlement - Number of housing
3.1.4. Public Facilities: -Public schools - Health facilities - Market - Religion facilities
3.1.5 Type and location of surrounding activities
3.2. Physical-Chemical Environmental Component
3.2.1. Climate: - Temperature - Humidity
3.2.2. Air Quality: - Wind direction and speed - Gases (CO, NOx, etc) - Dust - Noise
3.2.3. Physiography: - Topography - Geology (morphology, aquifer)
3.2.4. Hydrology
3.2.5. Hydro-oceanography
3.3. Biological Component
3.3.1. Land : - Type and bio-diversity of flora - Type and bio-diversity of fauna
3.3.2. Marine: - Type and bio-diversity of flora - Type and bio-diversity of fauna
3.4. Socio-economic and culture components
3.4.1. Demography: - Total population - Population density - Population growth - Population structure - Total population affected by the plan
3.4.2. Socio Economic: - Average income - Jobs - Employment opportunity
3.4.3. Socio Culture: - Culture value - Culture conflict - Culture diversity - Public facilities affected by the plan - Perception of community on the plan

2. Institutions for impact mitigation and impact monitoring

Avtivities	Institution Proposed
Executing institution for impact mitigation	Initiator
Monitoring institution for impact mitigation	- EIA provincial/district offices - District Industry office - Environmental research centre - NGO
Related institution for impact mitigation measures	- District Health office - District Industry office
Executing institution for environmental impact monitoring	Initiator
Monitoring Institution for environmental impact monitoring	- EIA provincial/district offices - District Industry office - Environmental research centre - NGO
Related Institution for environmental impact monitoring	- District Health office - District Industry office

APPENDIX III

Quality of 10 EISs reviewed: Summary

	10 reports								*Sea Communication reports (n=5)*								*Industry reports (n=5)*							
	Review topics				**Total weighted ranks**				**Review topics**				**Total weighted ranks**				**Review topics**				**Total weighted ranks**			
	10 reports		**4 cases**		**10 reports**		**4 cases**		**5 reports**		**2 cases**		**5 reports**		**2 cases**		**5 reports**		**2 cases**		**5 reports**		**2 cases**	
Review areas	score	rank	score	rank	score	rank	score	rank	score	rank	score	rank	score	rank	score	rank	score	rank	score	rank	score	rank	score	rank
0 Overall Assessment	**D**		**D**		**22**		**31**		**C**		**B**		19		22		**E**		**D**		**10**		**12**	
1. General description	D	R1	D	R1	23	R1	32	R1	B	R1	B	R1	35	R1	42	R1	D	R1	D	R1	20	R1	22	R1
1.1 Description of the Plan	C	1	D	1	9	1	11	2	B	3	B	3	32	1	38	1	D	1	D	2	22	1	24	1
1.2 The effected environment	D	2	D	2	8	2	11	1	B	1	B	1	12	2	15	2	D	2	C	1	7	2	8	2
1.3 Baseline conditions	E	3	E	3	8	2	15	3	B	1	B	1	13	2	15	2	E	3	E	3	5	3	5	3
2 Evaluation of key impacts	D	R1	D	R1	14	R4	16	R4	C	R3	B	R2	13	R3	17	R3	E	R1	D	R1	8	R3	9	R3
2.1 Scoping & env't appraisals	D	1	D	1	22	1	26	1	B	2	B	1	18	2	24	2	D	1	D	1	11	2	13	1
2.2 Description of key impacts	D	1	D	1	14	2	18	2	B	1	B	1	29	1	36	1	E	3	D	1	15	1	18	1
2.3 Assessment of impacts	E	1	D	1	7	4	7	4	C	3	B	3	17	3	23	2	D	1	D	1	11	2	13	3
2.4 Sustainability of the Plan	D	4	D	4	12	2	15	3	D	4	C	4	7	4	8	4	E	4	D	4	4	4	5	4
3. Mitigation and Monitoring	D	R1	D	R1	19	R2	34	R1	B	R2	B	R2	20	R2	21	R2	E	R1	D	R1	10	R2	13	R2
3.1 Alternative measures	D	1	D	1	6	2	25	1	C	4	B	2	30	1	35	1	E	4	D	3	12	1	23	1
3.2 Mitigation measures	D	3	D	3	10	1	16	2	C	3	B	2	22	2	23	2	E	3	D	1	11	2	13	2
3.3. Monitoring measures	E	3	D	3	3	4	5	3	B	1	A	1	15	3	17	3	D	1	D	3	8	3	8	3
3.4 Recommendations	D	2	D	2	7	2	8	4	C	2	C	4	4	4	4	4	D	1	D	1	3	4	3	4
4.Communication of results	D	R1	D	R1	14	R3	20	R3	C	R3	C	R4	9	R4	10	R4	E	R4	D	R4	5	R4	6	R4
4.1 Lay outs	D	2	D	2	21	1	25	1	B	3	B	1	24	2	28	1	D	2	D	2	14	2	16	2
4.2 Presentation	C	1	C	1	18	2	24	1	B	1	A	4	31	1	36	1	D	1	C	1	20	1	25	1
4.3 Uncertainties	D	3	D	4	11	3	11	3	D	5	D	4	13	4	14	4	E	3	D	4	10	2	11	2
4.4 Emphasis	D	3	D	2	7	4	10	3	B	1	B	4	18	2	21	3	E	3	D	2	8	2	10	2
4.5 Consultation	D	3	E	4	3	5	3	4	C	4	C	2	8	4	10	4	E	3	D	4	5	5	5	5
4.6 Non technical summary	F	6	F	5	0	6	0	5	D	5	D	3	3	5	3	5	F	6	F	5	1	5	1	5

Note: A, B, C: quality of report above the acceptance level
D, E, F: quality of report below the acceptance level

APPENDIX IV

Review Criteria for Terms of Reference (ToR) for the production of coastal EIS

Sources:

- Updated from EIS review criteria as in Occasional paper no 5 (Lee *et al* 1999),
- Adjusted to the Indonesian technical guidelines for the Sea Communications and the Industry sectors,
- Results of research on the EIS review process

A. Collation Sheet

1. **Assessment Symbols:** Use the following symbols when completing the collation sheet below.

Symbol	Explanation
A	Generally well performed, no important tasks left incomplete.
B	Generally satisfactory and complete only minors omissions and inadequacies.
C	Can be considered just satisfactory, despite omissions and/or inadequacies.
D	Parts are well attempted but must, as a whole, be considered just unsatisfactory because of omissions or inadequacies.
E	Not satisfactory, significant omissions or inadequacies.
F	Very unsatisfactory, important task (s) poorly done or not attempted.
NA	Not applicable. The Review topic is not applicable or it is irrelevant in the context of these environmental appraisal reports.

2. Scoring procedure

a. Use of symbols (A to NA) is only applied to the questions with three digits (example 1.1.1 to 1.1.6).
b. Change the symbol to number to get an average grade. (A=6; B=5; C=4; D=3; E=2; F=1, and NA=0).
c. Grade for point 1.1 is an average of grades from 1.1.1 to 1.1.6 (totals of grades divided by 6).
d. Grade for point 1 is an average of grades from 1.1 to 1.3. Grade for point 2 is an average of grades from 2.1 to 2.4. Etc.
e. Average grade for overall quality is an average of grades from 1 to 2 (total grades from 1 to 2 divided by 2).
f. Final grade for the overall assessment is the average grade in symbol value (6=A; 5=B; 4=C; 3=D; 2=E; 1=F; and 0=NA).

3. Collation sheet

Overall assessment ....

1 ....

1.1 ....

1.1.1 ....

1.1.2 ....

1.1.3 ....

1.1.4 ....

1.1.5 ....

1.1.6 ....

1.2 ....

1.2.1 ....

1.2.2 ....

1.3 ....

1.3.1 ....

1.3.2 ....

1.3.3 ....

2 ....

2.1 ....

2.1.1 ....

2.1.2 ....

2.1.3 ....

2.1.4 ....

2.2 ....

2.2.1 ....

2.2.2 ....

2.3 ....

2.3.1 ....

2.3.2 ....

2.3.3 ....

2.4 ....

2.4.1 ....

2.4.2 ....

2.4.3 ....

B. List of review topics for ToR of the EIS

1. Description of the plan, the affected environment and the baseline information

1.1. Description of the Plan

1.1.1 Are the type, purpose and lifetime of the plan, its stage in the plan preparation process and any future stages clearly explained?

1.1.2 Are related land use and sectoral plans identified?

1.1.3 Are the plan's main socio-economic, environmental and/or sustainability objectives clearly stated?

1.1.4 Are the plan's main policies and proposals, and their aims described?

1.1.5 Is the location and extent of the main areas allocated in the plan for different types of development indicated?

1.1.6 Are international or national environmental protection objectives (including objectives established in related plans) considered?

1.2. The affected environment

1.2.1 Is the local environment likely to be affected by the plan identified and described (by narrative description and/or by a map), including areas extending beyond the plan area, such as a catchment area?

1.2.2 Are components of the wider environment likely to be affected by the plan identified?

1.3. Baseline conditions

1.3.1 Is the local environmental stock described? Particular reference should be given to key assets and renewable and non renewable resources

1.3.2 Is the local environmental quality described? Existing environmental problems and pressures on the environment should be described, including estimates of waste production, pollution levels, and other development pressures on the environment

1.3.3 Are the baseline conditions described?

2. Identification and evaluation of key impacts

2.1. Scooping of the environmental appraisal

2.1.1 Are potentially significant policies, proposals and their impacts, including those of alternatives, identified using a systematic methodology?

2.1.2 Are environmental/sustainability indicators established and justified to assist in impact identification?

2.1.3 Are potentially significant impacts on the following environmental receptors: human beings; flora and fauna; soil; water; air; climate; landscape; material assets; and cultural heritage and interactions between them identified?

2.1.4 Are alternatives for achieving the plan's objectives or policies identified and is the reason for selecting these for further study given?

2.2. Describing key impacts

2.2.1. Is a description of the key impacts of the plan's policies and proposals and its alternatives, as identified at the scooping stage, given? A description should be given of any direct and indirect; cumulative; permanent and temporary; positive, negative or uncertain; short and long-term (quantified, where possible); impacts of the plan's policies and proposals on the local and wider environment.

2.2.2. Are the types of future changes to environmental media and receptors (as identified in 1.2) described?

2.3. Assessment of impacts

2.3.1. Is impact magnitude predicted, either in quantitative or qualitative terms?

2.3.2. Is impact significance assessed, taking into consideration where relevant: impact magnitude; impact locations; impact duration; opinions of affected parties/experts; environmental/sustainability criteria; the precautionary principle; international and national environmental protection objectives?

2.3.3. Is the methodology used to predict impact magnitude and significance described and justified? Any value judgements used should be explicitly stated.

2.4. Appraising the sustainability of the plan

2.4.1 Does the Report assess the sustainability of the plan in the local and wider context?

2.4.2 Does the Report established and justify specific criteria for evaluating the sustainability of the plan?

2.4.3 Does the Report assess the compliance of the plan to national or Local Authority sustainability strategies, e.g. Local Agenda 21?

APPENDIX V

EIS Review Criteria

Sources:

- Updated from EIS review criteria as in Occasional paper no 5 (Lee *et al* 1999),
- Adjusted to the Indonesian technical guidelines for the Sea Communications and the Industry sectors,
- Results of research on the EIS review process.

<u>A. Collation Sheet</u>

1. **Assessment Symbols:** Use the following symbols when completing the collation sheet below.

Symbol	Explanation
A	Generally well performed, no important tasks left incomplete.
B	Generally satisfactory and complete only minors omissions and inadequacies.
C	Can be considered just satisfactory, despite omissions and/or inadequacies.
D	Parts are well attempted but must, as a whole, be considered just unsatisfactory because of omissions or inadequacies.
E	Not satisfactory, significant omissions or inadequacies.
F	Very unsatisfactory, important task (s) poorly done or not attempted.
NA	Not applicable. The Review topic is not applicable or it is irrelevant in the context of these environmental appraisal reports.

2. Scoring procedure

a. Use of symbols (A to NA) is only applied to the questions with three digits (example 1.1.1 to 1.1.6).
b. Change the symbol to number to get an average grade. (A=6; B=5; C=4; D=3; E=2; F=1, and NA=0).
c. Grade for point 1.1 is an average of grades from 1.1.1 to 1.1.6 (totals of grades divided by 6).
d. Grade for point 1 is an average of grades from 1.1 to 1.3. Grade for point 2 is an average of grades from 2.1 to 2.4. Etc.
e. Average grade for overall quality is an average of grades from 1 to 2 (total grades from 1 to 2 divided by 2).
f. Final grade for the overall assessment is the average grade in symbol value (6=A; 5=B; 4=C; 3=D; 2=E; 1=F; and 0=NA).

2. Collation sheet

Overall assessment …………

1 ……..
1.1 ……..
1.1.1 ……..
1.1.2 ……..
1.1.3 ……..
1.1.4 ……..
1.1.5 ……..
1.1.6 ……..

1.2 ……..
1.2.1 ……..
1.2.2 ……..

1.3 …….
1.3.1 …….
1.3.2 …….
1.3.3 …….

2 ……..
2.1 ……..
2.1.1 ……..
2.1.2 ……..
2.1.3 ……..
2.1.4 ……..

2.2 ………
2.2.1 ………
2.2.2 ………

2.3 ………
2.3.1 ………
2.3.2 ………
2.3.3 ………

2.4 ………
2.4.1 ………
2.4.2 ………
2.4.3 ………

3 ………
3.1 ………
3.1.1 ………
3.1.2 ………

3.2 ………
3.2.1 ………
3.2.2 ………
3.2.3 ………

3.3 ………
3.3.1 ………
3.3.2 ………
3.3.3 ………

3.4 ………
3.4.1 ………
3.4.2 ………

4 ……….
4.1 ……….
4.1.1 ……….
4.1.2 ……….
4.1.3 ……….
4.1.4 ……….

4.2 ……….
4.2.1 ……….
4.2.2 ……….
4.2.3 ……….

4.3 ……….
4.3.1 ……….
4.3.2 ……….

4.4 ……….
4.4.1 ……….
4.4.2 ……….

4.5 ……….
4.5.1 ……….
4.5.2 ……….

4.6 ……….
4.6.1 ……….
4.6.2 ……….

B. List of review topics

1. Description of the plan, the affected environment and the baseline conditions

1.1. Description of the Plan

1.1.1. Are the type, purpose and lifetime of the plan, its stage in the plan preparation process and any future stages clearly explained?
1.1.2. Are related land use and sectoral plans identified?
1.1.3. Are the plan's main socio-economic, environmental and/or sustainability objectives clearly stated?
1.1.4. Are the plan's main policies and proposals and their aims described?
1.1.5. Is the location and extent of the main areas allocated in the plan for different types of development indicated?
1.1.6. Are international or national environmental protection objectives (including objectives established in related plans) considered?

1.2. The affected environment

1.2.1. Is the local environment likely to be affected by the plan identified and described (by narrative description and/or by a map), including areas extending beyond the plan area, such as a catchment area?
1.2.2. Are components of the wider environment likely to be affected by the plan identified?

1.3. Baseline conditions

1.3.1. Is the local environmental stock described? Particular reference should be given to key assets and renewable and non renewable resources.
1.3.2. Is the local environmental quality described? Existing environmental problems and pressures on the environment should be described, including estimates of waste production, pollution levels, and other development pressures on the environment.
1.3.2. Are the baseline conditions described?

2. Identification and evaluation of key impacts

2.1 Scooping of the environmental appraisal

2.1.1 Are potentially significant policies, proposals and their impacts, including those of alternatives, identified using a systematic methodology?
2.1.2 Are environmental/sustainability indicators established and justified to assist in impact identification?
2.1.3 Are potentially significant impacts on the following environmental receptors: human beings; flora and fauna; soil; water; air; climate; landscape; material assets; and cultural heritage and interactions between them identified?
2.1.4 Are alternatives for achieving the plan's objectives or policies identified and is the reason for selecting these for further study given?

2.2. Describing key impacts

2.2.1 Is a description of the key impacts of the plan's policies and proposals and its alternatives, as identified at the scooping stage, given? A description should be given of any direct and indirect; cumulative; permanent and temporary; positive, negative or uncertain; short and long-term (quantified, where possible); impacts of the plan's policies and proposals on the local and wider environment.

2.2.2 Are the types of future changes to environmental media and receptors (as identified in 1.2) described?

2.3. Assessment of impacts

2.3.1 Is impact magnitude predicted, either in quantitative or qualitative terms?

2.3.2 Is impact significance assessed, taking into consideration where relevant: impact magnitude; impact locations; impact duration; opinions of affected parties/experts; environmental/sustainability criteria; the precautionary principle; international and national environmental protection objectives?

2.3.3 Is the methodology used to predict impact magnitude and significance described and justified? Any value judgements used should be explicitly stated.

2.4. Appraising the sustainability of the plan

2.4.1 Does the Report assess the sustainability of the plan in the local and wider context?

2.4.2 Does the Report established and justify specific criteria for evaluating the sustainability of the plan?

2.4.3 Does the Report assess the compliance of the plan to national or Local Authority sustainability strategies, e.g. Local Agenda 21?

3. Alternatives, Mitigation measures, Monitoring, and Recommendations

3.1. Alternatives

3.1.1 Are alternatives, considered at previous and present stages of the plan making process, described and evaluated and the reasons for any final choices given? Reasons for not adopting alternatives should also be given. Alternatives may relate to: objectives; policies and proposals; location strategies and land use types.

3.1.2 Is the significance of the predicted environmental impacts used in justifying the choices between alternatives?

3.2. Mitigation measures

3.2.1 Are mitigating measures proposed to prevent, reduce or offset the significant adverse impacts of implementing the plan's policies and proposals on the environment?

3.2.2 Is the anticipated effectiveness of the proposed mitigation measures indicated?

3.2.3 Is the commitment to, and responsibilities for, mitigation measures stated?

3.3 Monitoring and review

3.3.1 Are monitoring arrangements proposed to check the environmental impacts resulting from the implementation of the plan, and their conformity with the predictions within the Report?

3.3.2 Are there provisions to review the plan on a regular basis to ensure that any unexpected environmental impacts are identified and taken into account in plan revisions?

3.3.3 Are environmental/sustainability indicators established and justified to assist in monitoring activity?

3.3.4 Is the commitment to and responsibilities for monitoring and review stated?

3.4 Recommendations

3.4.1 Does the Report contain recommendations concerning the contents of the plan, e.g. amending and introducing new policies or proposals and/or on the final selection of alternatives?

3.4.2 Are recommendations made for the further investigation of activities arising from the plan within the project-level environmental assessment process?

4. Communication of Results

4.1. Lay out

4.1.1 Is information logically arranged in sections or chapters? The structure of the appraisal should be indicated in a table of contents and summarised in its introduction.

4.1.2 Are the predicted environmental impacts referenced to the policies giving rise to them?

4.1.3 Do chapters and other sections of the Report, unless very short, contain summarised outlining their main findings and conclusions?

4.1.4 Where data or material from external sources are introduced, is the original source acknowledged at that point in the text? Such data and material should be adequately referenced.

4.2. Presentation

4.2.1 Is information presented so as to be comprehensible to the non-specialist? Tables, graphs, sketch maps and other devices should be used as appropriate.

4.2.2 Is obscure of language avoided? Acronyms and initials should be defined.

4.2.3 To what extent is the Report presented as a self-contained document?

4.2.4 If important data and material are located in appendices, are they also summarised, presented and discussed in the main body of the text?

4.3. Uncertainties

4.3.1 Are uncertainties and other limitations regarding information, data and methodologies acknowledged?

4.3.2 Does the Report explain and justify how these uncertainties and limitations have been handled within the environmental appraisal?

4.4. Emphasis

4.4.1 Are both significance adverse and beneficial environmental impacts given their due emphasis? The significance of adverse impacts should not be disguised by empty or imprecise phrases.

4.4.2 Is the information in the Report presented without bias? The report should not lobby for a particular point of view toward the plan and its likely environmental consequences.

4.5. Consultation.

4.5.1 Is information presented in the Report on any consultation exercises undertaken, during the environmental appraisal, with the environmental authorities, NGOs, the general public and other interested parties in the development plan process?

4.5.2 Are the opinions they expressed summarised and taken into account in the Report?

4.6. Non-technical summary.

4.6.1 Is there a non-technical summary of the environmental appraisal Report? This should include a brief description of the plan, its main objectives and alternatives considered.

4.6.2 Does the summary cover all major Review Topics and issues in the Report and its principal findings and recommendations?

APPENDIX VI

List of Respondents and Experts

(Research conducted in the period 1999-2000)

For the review of the EIA guidelines and the EIS:
- Ir. Nana Sudiana
- Drs. Bambang Sudjatmiko

For the weighted scales of the EIS review:
- Ir. Kustaman, a representative of the District Government of Serang who is involved in the EIS review processes.
- Ir. Yani, a representative of the District Government of Serang who is dealing with environmental monitoring

For the review of environmental monitoring:
- Ir. Irawan Noor and Ir. M. Ridwan from the District Planning Office of Serang.
- Ir. Anang Maulana and Ir. Yani from the District Environment Office of Serang.

For the assessment on perception of technical teams on the EIA:
- Ir. Hariyanto Sagala, the secretary of technical team of the Industry office
- Ir. Ninuk, a member of technical team of the Industry office
- Drs. Amien Suwarto, MBA, the secretary of technical team of the Sea Communications office
- Ir. Hary Budiarto, a member of technical team of the Sea Communications office
- Ir. Suryo Adiwibowo, an advisor to the State Minister of Environment.

APPENDIX VII

A quick tour around Banten bay Information System

A Marine Ecosystem Information System to support EIA (AMDAL) and Integrated Coastal Zone Management in Banten Bay, Indonesia

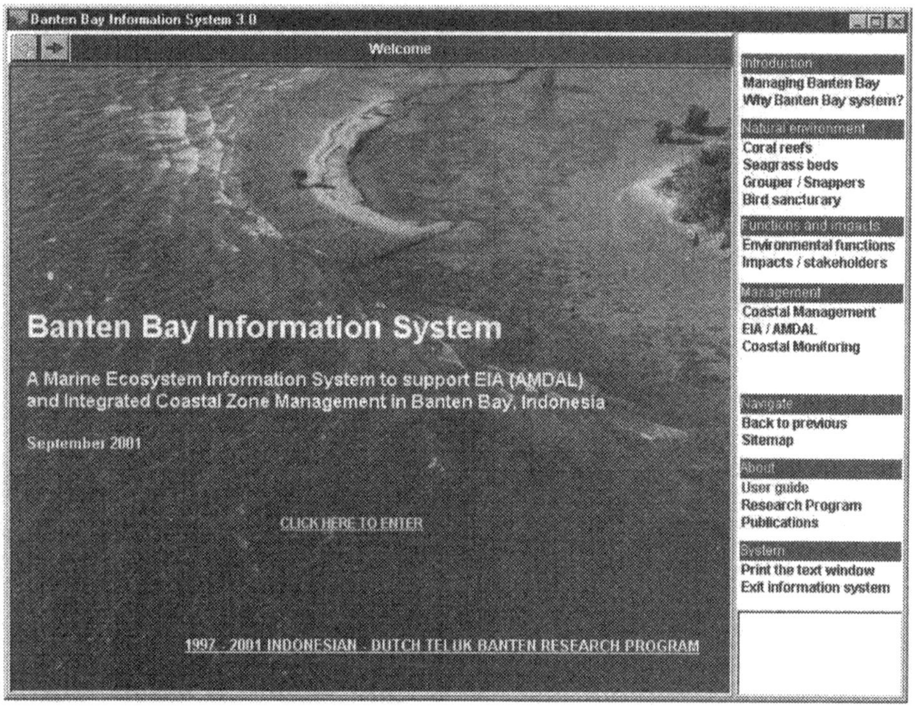

September 2001
IHE, Delft / BPPT, Jakarta

1997-2001 Indonesian - Dutch Teluk Banten Research Programme

ACKNOWLEDGEMENT

Banten Bay Information System was developed in the framework of the Teluk Banten Research Programme, an Indonesian-Dutch research programme on Integrated Coastal Zone Management (1997 - 2001). The Royal Netherlands Academy of Arts and Sciences (KNAW), the Foundation for the Advancement of Tropical Research (WOTRO), The Netherlands and the Indonesian Government fund the programme.

Technical requirements of Banten Bay Information System

Banten Bay Information System runs under Windows 95/98/NT. An IBM compatible 486 or Pentium computer is recommended to run the system. For the multi-media part of the system a Pentium is recommended. Installation of Banten Bay Information System requires about 80 MB of free disk space. The program has been designed for a screen resolution of 1024 by 768.

To run the information system, Microsoft Excel 97 or 2000 is needed. In addition, Internet Explorer 4.0 or higher, Windows Media Player, Adobe Acrobat Reader, and MapObjects LT are needed. These freeware programs are available from the CD-Rom.

Further information

For information please contact Wim Douven or Jetze Heun, IHE, Delft (wjd@ihe.nl; (31)-15-215.1712) or Dwi Abad Tiwi, BPPT, Jakarta (dwiabadtiwi@yahoo.com; (62)-21-3169652).

WHY BANTEN BAY INFORMATION SYSTEM?

The Banten Bay coastal zone is located on the north-west coast of Java. It is a dynamic area with a fast growing population and many large-scale industrial developments. These developments have important economic and social benefits for the local as well as national population. They, however, also pose serious threats to the marine environment, including threats to ecosystems as coral reefs, seagrass beds, groupers and snappers, and the water bird Nature Reserve of Pulau Dua. The environmental functions of these ecosystems are valued differently by the various stakeholder groups. Marine environmental functions most recognised are the ones providing direct benefits such as fish production. Other functions, such as coastal protection and natural beauty, have less visible benefits and are less taken into account in decision-making.

The challenge of the planners and stakeholders is how to develop coastal and marine resources for the maximum benefit of the district and at the same time, to maintain resources and ecosystems at sustainable levels. The District Planning Office (Bappeda) plays a central role in taking up this challenge. The recently approved laws, e.g. on the regional autonomy, have strengthened this role. To be able to carry out this role information on the status and the functioning of marine ecosystems is essential. Banten Bay Information System contains:

- information (text, causal diagrams, databases, GIS, photos) on the bay, including coral reefs, seagrass beds, grouper and snapper populations, the water birds of Pulau Dua Nature Reserve, and human activities putting pressure on these ecosystems;
- information on environmental functions of the ecosystems and their values;
- a description of the use of coastal zone indicators, including scientific background information, to monitor the ecosystems; and
- guidance on how this information can support Integrated Coastal Zone Management and improve Environmental Impact Assessments (AMDAL).

A QUICK TOUR AROUND THE SYSTEM

Structure and use of the system

The structure of the Banten Bay Information System is shown in Figure 1. The diagram shows the main components of the information system: the natural environment, its functions, the stakeholders who make use of these functions, and the management of uses and impacts. With respect to the natural environment the focus of the information system is on coral reefs, seagrass beds, groupers and snappers and Pulua Dua Nature Reserve. The focus of management on the management of impacts by means of Environmental Impact Assessments (EIA). By clicking on the components in the diagram the different information system modules can be accessed. The information system modules can also be accessed by means of the menu (in blue headings) at the right. Information in Banten Bay Information System is available in various forms, such as text, diagrams, tables, maps, photos and video. Most information will be displayed in the large window at the left. More detailed information can be obtained by following a hyperlink (like the internet). Hyperlinks can be either words (blue and underlined) or so-called hotlinks (as used in the multi-media map; Figure 2). The information stored in Excel databases and GIS can be accessed by clicking buttons (see e.g. Figure 8).

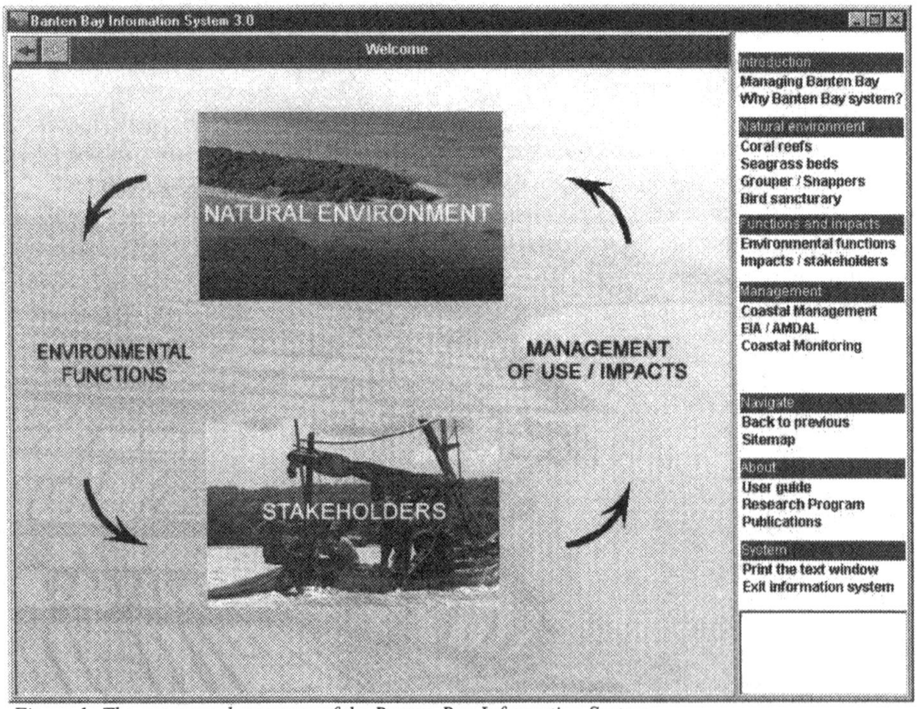

Figure 1: The conceptual structure of the Banten Bay Information System.

The Introduction Module

The introduction module has two entrances in the menu: 'Managing Banten Bay' and 'Why Banten Bay Information System?'. The first entrance gives introductory information about Banten Bay, its natural environment, the main activities and related issues, as well as the role of coastal zone management. In several ways the user can get an impression of the Banten Bay coastal zone. Figure 2 shows the multi-media map, and Figure 3 the 'GIS view' that offers the possibility to view maps of various themes.

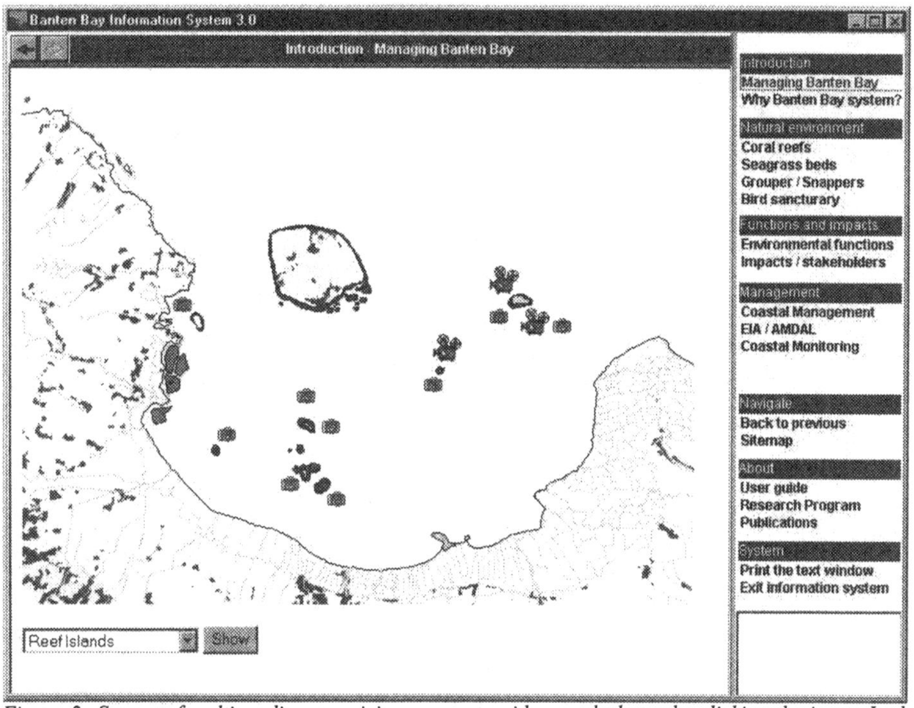

Figure 2: Screen of multi-media map giving access to videos and photos by clicking the icons. In the example shown the theme ' Reef islands' is selected.

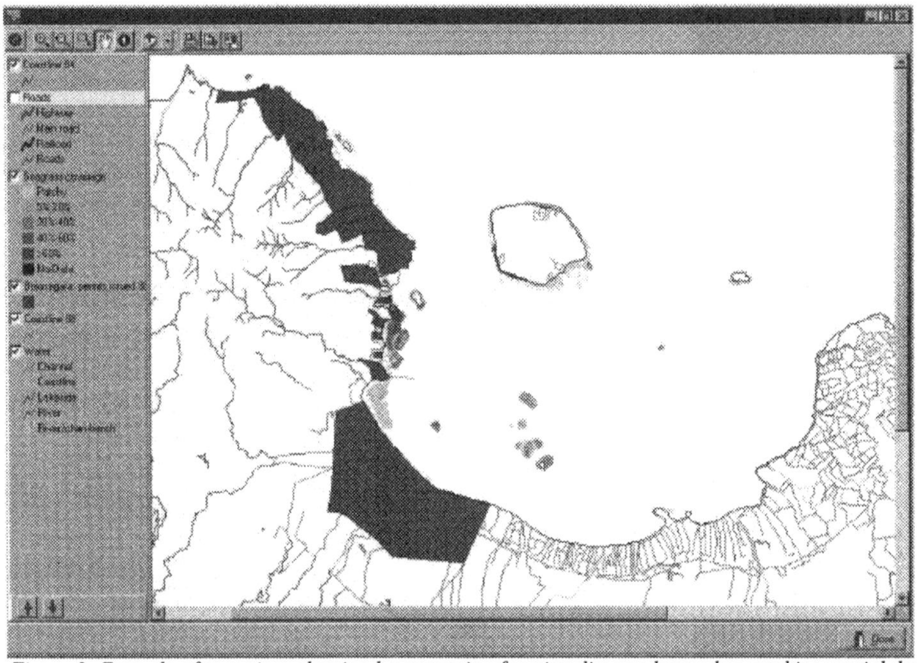

Figure 3: Example of map view, showing how mapping functionality can be used to combine spatial data layers, and investigate possible impacts.

The Stakeholder / Impact Module

This module starts with a map showing the different regions in the Banten Bay coastal zone (Figure 4). After clicking one of the regions a table will be presented listing the human activities and expected economic and environmental impacts of these activities. The user has the option to access more detailed data by clicking on hyperlinks in the table. For instance, after clicking on an activity a listing is presented of the related stakeholder(groups); see Figure 5 for an example of the stakeholders related to the fishery activity. The stakeholder data is stored in an Excel database and can be easily updated.

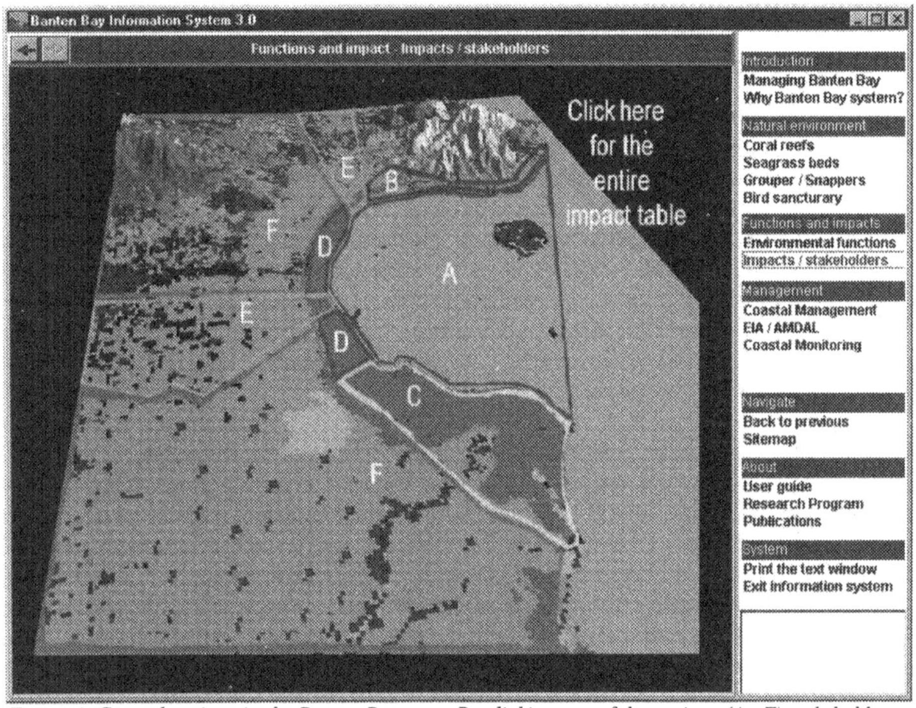

Figure 4: Coastal regions in the Banten Bay area. By clicking one of the regions (A - F) stakeholder and impact information will be displayed.

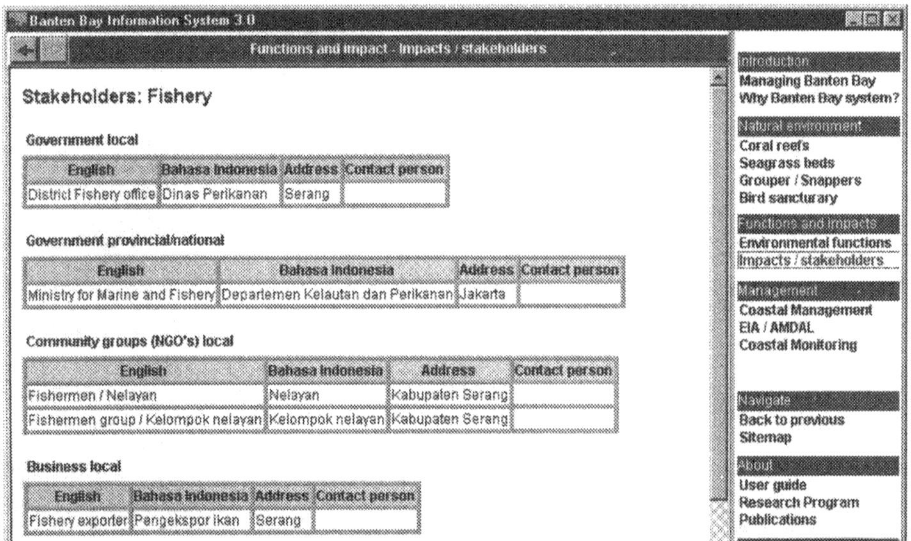

Figure 5: Banten Bay stakeholders related to the fishery activity.

The Natural Environment Module

The Natural Environment Module gives information on the bay, including coral reefs, seagrass beds, grouper and snapper populations, the water birds of Pulau Dua Nature Reserve, and human activities putting pressure on these ecosystems. Further, a description is given of the use of coastal zone indicators, including scientific background information, to monitor these ecosystems.

Figure 6 shows as an example the screen that will be displayed after selecting the seagrass ecosystem. The item 'The Banten Bay seagrass' gives introductory information about the seagrass with maps and photo's. The item 'Description of the seagrass system' gives information about functional relationships between system components. When clicking on 'The physical environment' a general description is given of the physical condition of the bay (see Figure 7).

After selecting 'Indicators for monitoring' a screen is displayed showing a diagram in the upper half of the text window. Figure 8 shows this screen of the coral reefs part of the module. The boxes of the diagram represent the different human pressure and natural state indicators of the reef system. After clicking on a box, information about that indicator will be displayed in the lower half of the text window. Dependent on the information available for this indicator the user can further access time series, map information, background information about the baseline data set, monitoring guidelines, and threshold information.

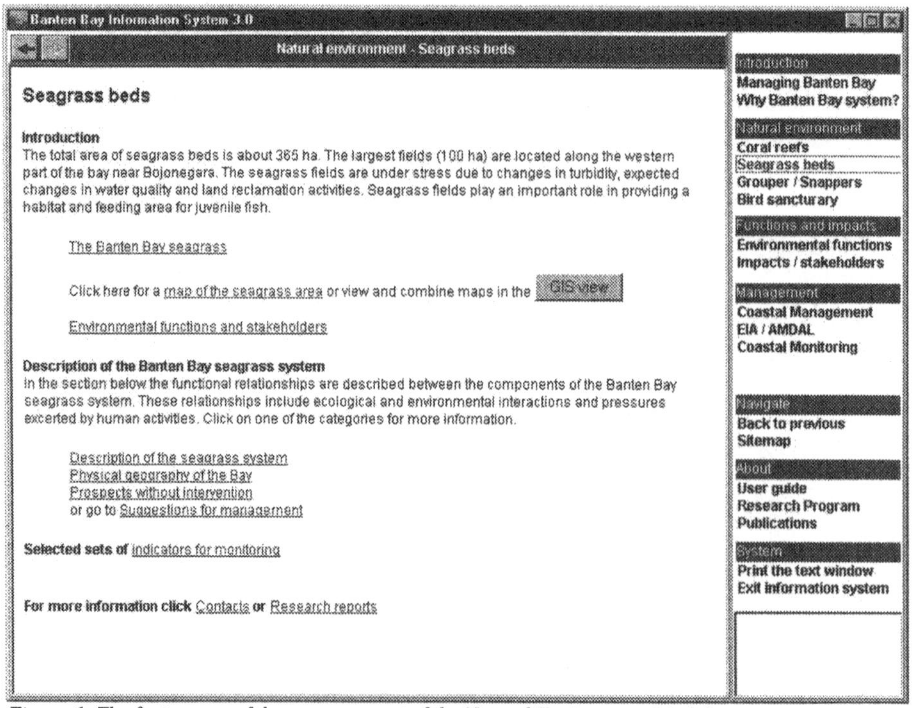

Figure 6: The first screen of the seagrass part of the Natural Environment Module.

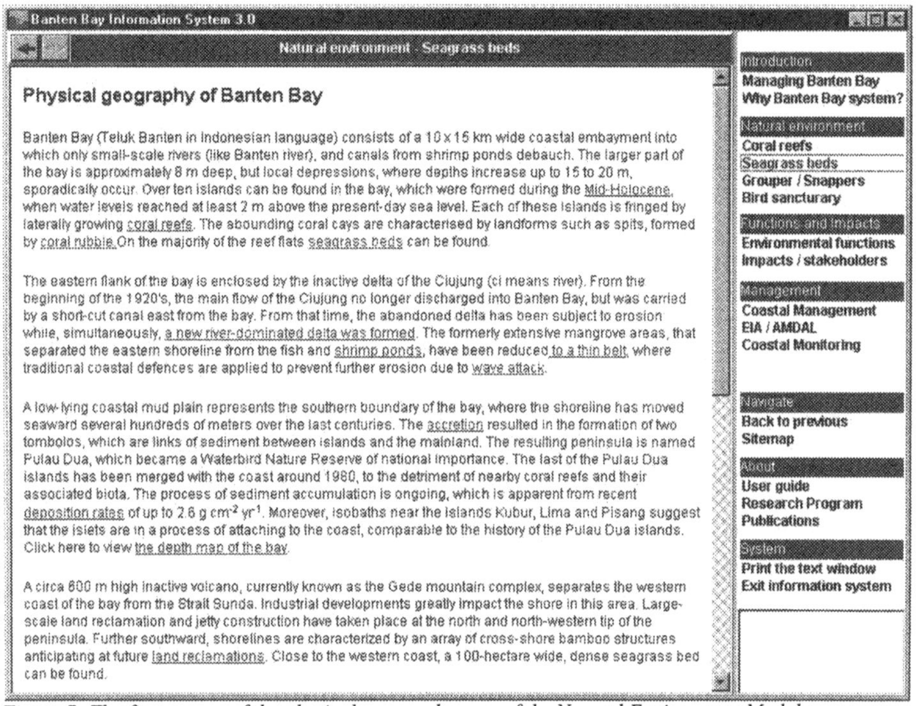

Figure 7: The first screen of the physical geography part of the Natural Environment Module.

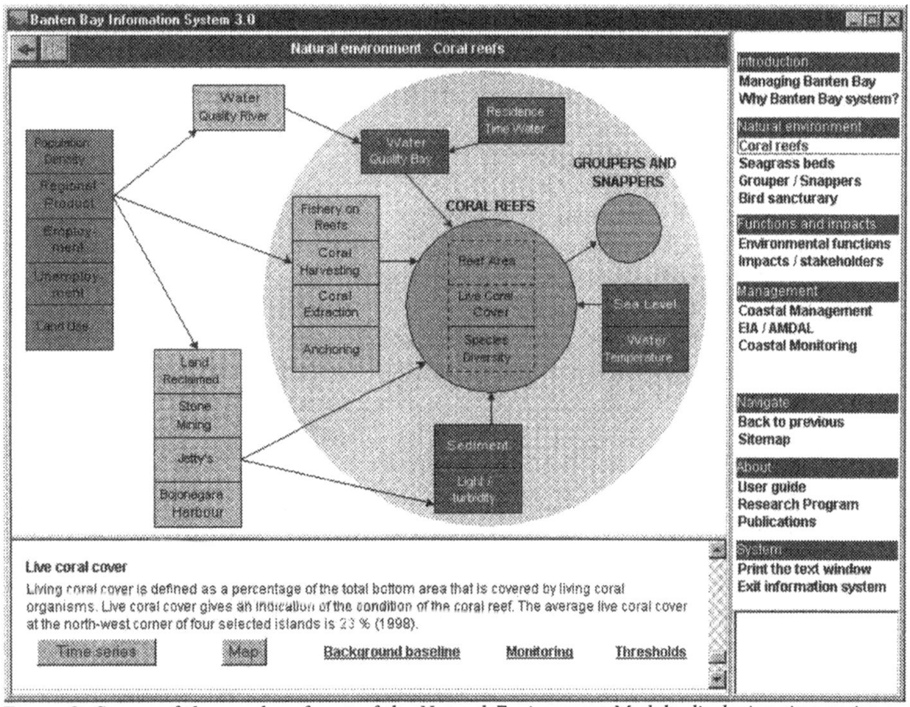

Figure 8: Screen of the coral reef part of the Natural Environment Module displaying time series and spatial information of the reef indicator ' Live coral cover'.

The EIA / AMDAL Module

This module supports the Environmental Impact Assessment procedures of planned developments in the Banten Bay coastal zone. The diagram of Figure 9 shows the Indonesian EIA procedure. The various steps are clickable and will lead the user to general information about this phase, to suggestions how to carry out the phase, and to relevant modules, such as the 'Natural environment' and 'Functions and impacts' module. Below this will be illustrated for some EIA stages.

The 'Scoping' phase

The scoping phase is by many considered as a crucial phase as it is here where the issues to be addressed in the EIA study considered will be explored and listed. Questions, like 'What impacts to take into account in this EIA study?, and 'What stakeholders to involve in the EIA process?' are relevant in this phase. Figure 10 shows the screen that explains how the system supports this phase, with links to relevant system modules.

The 'Baseline study' phase

In a baseline study information on the current status of the environment likely to be affected by the development is collected. The Natural Environment Module of Banten Bay Information System contains a scientific baseline of the four ecosystem components and related human activities and physical elements (see e.g. Figures 7 and 8). In several ways the baseline stored in the information system supports the development of an EIA baseline:

- it gives general information about the natural marine situation;
- it gives background values of marine ecological conditions;
- it is a reference in the selection of marine ecological indicators to be taken into account in the EIA baseline; and
- it gives information about research institutes, emails researchers, background documents, etc.

The 'Review / approval of EIS' phase

Based on the Ministerial decree for the guidance of the EIS review, on the EIA technical guidelines for the Sea Communication and the Industry sectors, and based on research results, an EIS review package has been compiled and linked to the EIA / Amdal Module. The package consists of guidance on how to use the package, a list of review criteria, and a collation sheet to record the findings. The aims of the package are to offer the EIS reviewer a more simple review procedure and to obtain more transparent review results.

The Coastal Zone Monitoring Module

This module gives suggestions for monitoring the Banten Bay ecosystem. The sets of indicators stored in the information system can be used as a reference during the selection of indicators. These indicators will be described in a Coastal Zone Management plan or an Environmental Impact Statement (in case of EIA), and then monitored to see whether there are differences with the real situation after the development has taken place. On the basis of the results it can be decided whether adjustments are required. To give guidance in this process this module links the user to the 'Indicators for monitoring' part of the four ecosystem components (Natural Environment Module).

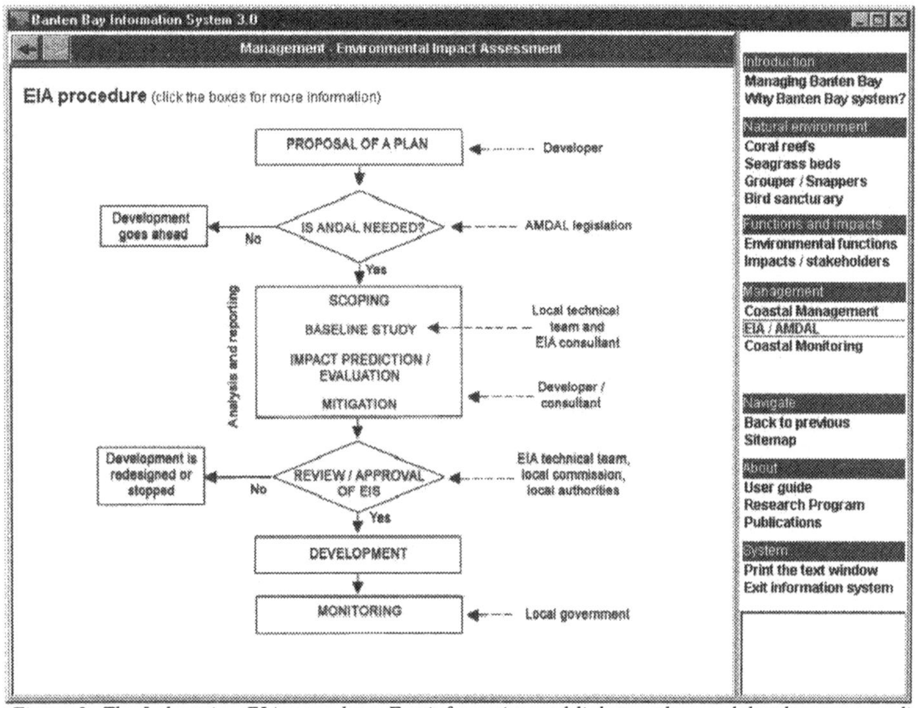

Figure 9: The Indonesian EIA procedure. For information and links to other modules the user can click on the EIA steps.

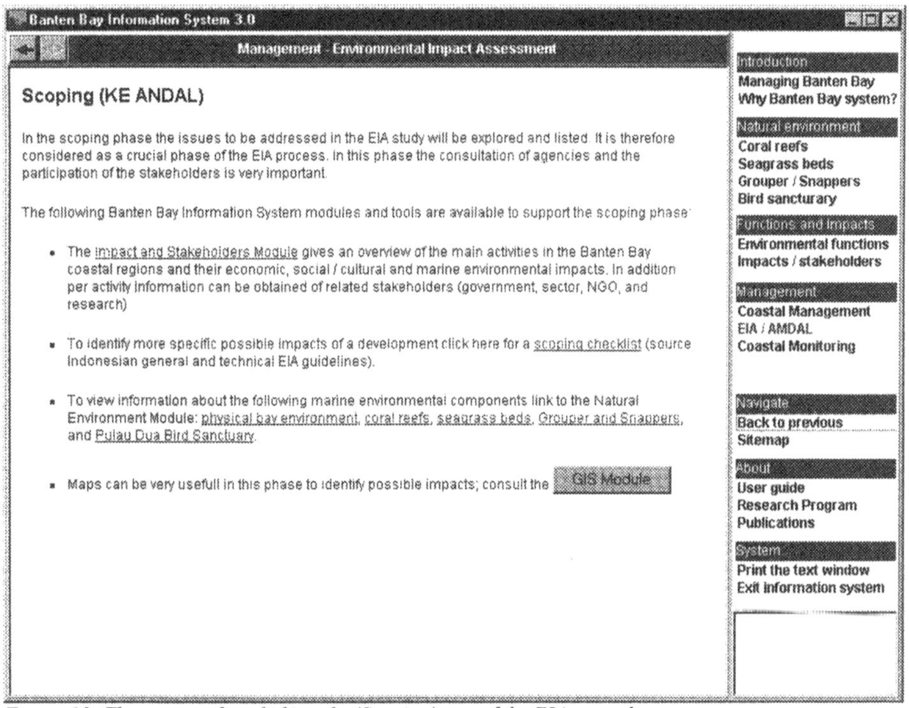

Figure 10: The screen after clicking the 'Scoping' step of the EIA procedure.

How to make further use of Banten Bay Information System?

Coastal zone management requires an integral approach that needs involvement of all relevant government agencies, institutes, sectors and other stakeholders. One of the pre-conditions for co-operation between these stakeholders is the integration and dissemination of information of the various themes related to the coastal zone. Banten Bay Information System can be considered as a first step (for a selected part of the coastal zone system) towards such an information platform where information and monitoring data from different sources (e.g. Kantor Statistik, PSDA, LIPI Jakarta, Teluk Banten Research Programme, etc..) is brought together in a coastal zone management context. Banten Bay Information System is designed in such a way that information can be updated and extended quite easily:

- text information can be added by e.g. linking new pages in a so-called html-editor (used for making internet pages);
- monitoring data can be added in the Excel database;
- stakeholder information can be added in the Excel database;
- data on information sources can be added in the Excel database; and
- maps can be added in the map view.

The system runs at this moment as a stand-alone system. In the future the system could be placed on the internet and then links can easily be made to external information sources.